T0321047

Plant Micronutrients

C.P. Sharma
Department of Botany
Lucknow University
Lucknow, India

Science Publishers
Enfield (NH)   Jersey   Plymouth

SCIENCE PUBLISHERS
An Imprint of Edenbridge Ltd., British Isles.
Post Office Box 699
Enfield, New Hampshire 03748
United States of America

Website: *http://www.scipub.net*

sales@scipub.net (marketing department)
editor@scipub.net (editorial department)
info@scipub.net (for all other enquiries)

Library of Congress Cataloging-in-Publication Data

Sharma, C.P. (Chandra Prakash), 1934-
  Plant micronutrients : roles, responses, and amelioration of deficiencies / C.P. Sharma.
    p. cm.
  Includes bibliographical references and index.
   ISBN 1-57808-416-4
   1. Plants--Nutrition. I. Title.

  QK867.S45 2006
  572'.42--dc22

                                              2006042252

ISBN 1-57808-416-4

Published by Science Publishers, NH, USA
An Imprint of Edenbridge Ltd.
Printed in India

Preface

Phenomenal progress has been made in our understanding of the several and diverse ways in which micronutrients alter or regulate the functioning of plants and produce a wide array of responses enabling quantitative and qualitative changes in plant yield. Developments on different aspects of micronutrients in plants have been reviewed and their implications discussed in several contemporary publications dealing with them, both collectively and individually. The author's resolve to add a resource on the biochemical and physiological roles of micronutrients in plants and the diverse ways in which they react to limitations in micronutrient supply, contributing to global constraints in crop production, came from his long stint as a teacher of plant nutrition at the Lucknow University. This need was augmented by spectacular developments in molecular biology. Immunocytochemical localization of specific proteins, functional analyses of phenotypic mutants, cloning and sequence analysis of genes and their expression in transgenics have provided new insight into the functional sites of micronutrients, their roles in macromolecular biosynthesis, and developmental processes and adaptive response to biotic and abiotic stresses. The author's long association with the All India Coordinated Research Project on Micronutrients of Soil and Plants, supported by the Indian Council of Agricultural Research, New Delhi, provided the impetus for adding a resource on the evaluation and amelioration of micronutrient deficiencies. These deficiencies account for a wide gap between the production potential of the major food crops and their productivity on low micronutrient soils and also necessitate their prompt management for sustained increase in crop yields. In keeping with these imperatives, the book has been presented in two parts. Part I deals with roles of micronutrients and the changes induced in plants in response to their inadequate supply. Part II addresses the factors contributing to micronutrients deficiency, their diagnosis, evaluation and management through soil - and plant - based approaches.

iv

The author is deeply indebted to late Professor SC Agarwala, who not only inducted him to plant nutrition research, but also provided strong motivation to prepare resource material for students opting for advanced courses on physiology and biochemistry of micronutrients. During the preparation of the book, the author received valuable help from Dr BD Nautiyal, Dr Nalini Pandey, Dr GC Pathak and Ms R Pandey, all from the Department of Botany, Lucknow University. The author would like to make particular mention of the support and helpful suggestions received in the making of the book from Dr Nalini Pandey and his son Dr Samir Sharma, faculty with the Lucknow University Department of Biochemistry. The author is grateful to Dr HK Jain, Director, Center for Science Writing, New Delhi, who not only encouraged him to work on the book but also provided financial support to defray part of the pre-publication costs. Thanks are also due to Mr Abhist Kumar for typing the first draft of the manuscript, which made difficult reading.

The author is conscious of the limitations in bringing out the book single handed, which might have left many deficiencies. Any criticism and suggestions that may help overcome the deficiencies will be greatly welcome.

Dec. 2005

C.P. Sharma
Department of Botany
Lucknow University
Lucknow

Contents

PART I

ROLES AND DEFICIENCY RESPONSES

Chapter 1

GENERAL

1.1 MICRONUTRIENTS AS ESSENTIAL PLANT NUTRIENTS

The autotrophic plants have an absolute requirement of the following 16 elements:

Carbon (C), Hydrogen (H), Oxygen (O), Nitrogen (N), Phosphorus (P), Potassium (K), Sulphur (S), Calcium (Ca), Magnesium (Mg), Iron (Fe), Manganese (Mn), Copper (Cu), Zinc (Zn), Molybdenum (Mo), Boron (B), Chlorine (Cl).

Of these 16 elements, first three—C, H and O—are derived from air and water. The remaining 13 elements are all derived from the soil-forming minerals, and so also known as **mineral nutrient elements** or simply **mineral nutrients**. They are all essential for plants in the sense that they satisfy the criteria of essentiality as laid down by Arnon and Stout (1939), which reads as:

(1) The plant must be unable to grow normally or complete its life cycle in the absence of the element.
(2) The element is specific and/or cannot be replaced by another element.
(3) The element plays a direct role in metabolism.

Based on their chemical (metallic or non metallic) nature, the form taken up by plants, biochemical functions and quantitative differences in functional requirements, the mineral nutrients have been divided into two or more categories. The distinction based on quantitative differences in functional requirements—**macronutrient elements** and the **micronutrient elements**, or for brevity just **macro-** and **micro-nutrients** (Loomis and Shull, 1937)—is most widely followed. Elements which, in solution cultures are required at a concentration exceeding 1 mg l^{-1}, often many times more viz. N, P, K, Ca, Mg or S, are classified as macronutrients; Those required in smaller quantities (<1 mg l^{-1}), viz., Mn, Cu, Zn, Mo, B, Cl are classified as micronutrients. Iron, which quantitatively behaves as a macronutrient, is for historical reasons, considered a micronutrient (Arnon, 1954).

Except for iron—which was known to be essential for plants even before von Sachs and William Knop formulated the nutrient solutions for soil-less culture of plants (in the 1860s)—unequivocal evidence for the essentiality of the micronutrients came after improvement in solution culture methods (Hewitt, 1966). The year of establishment of essentiality and the names of scientists credited with the discovery are shown in Table 1.1.

Table 1.1 Establishment of the essentiality of plant micronutrients

Micronutrient	Credit for definitive evidence of essentiality	Year
Manganese	J.S. Mc Hargue	1922
Boron	K. Warrington	1923
Zinc	AL Sommer and CB Lipman	1926
Copper	AL Sommer, CB Lipman and G. Mac Kinney	1939
Molybdenum	DI Arnon and PR Stout	1939
Chlorine	TC Broyer, AB Carlton, CM Johnson and PR Stout	1954

Cobalt and nickel, known to be essential for animals, have appeared to be good contenders for addition to the list. Cobalt has been shown to be essential for symbiotic nitrogen fixation by legumes (Ahmed and Evans, 1960; Dilworth et al. 1979) and also non legumes (Hewitt and Bond, 1966, Johnson et al. 1966). Cowles et al. (1969) showed cobalt to be essential for rhizobial growth. Role of cobalt in N_2 fixation is essentially attributed to its role as a cofactor of cobalamine (Vitamin b_6) which functions as a coenzyme involved in N_2 fixation and nodule growth (Dilworth et al. 1979; Licht et al. 1996; Jordan and Reichard, 1998). Wilson and Nicholas (1967) produced evidence of cobalt requirement for nodule forming legumes and wheat. Nickel is another element which has justified claims for its essentiality (Gerendas et al. 1999). Nickel is present in substantial concentrations in the xylem and phloem sap of plants. In the xylem sap, nickel is contained both in free ionic form (Ni^{2+}) and as chelate (Ni-citrate). Nickel has high mobility in phloem, which contributes to its translocation and accumulation in fruits and seeds. Wheat, barley and oats grown with nickel-free nutrient solution have been reported to develop visible symptoms such as intervenal chlorosis (Brown et al. 1987). Dixon et al. (1975) showed nickel to be a cofactor of jack bean urease. Plants grown with urea as the sole source of nitrogen showed absolute requirement of nickel for urease activity (Gordon et al. 1978; Gerendas and Sattelmacher, 1997). Claims for nickel as an essential plant nutrient have been reviewed by Gerendas et al. (1999).

1.2 MICRONUTRIENTS AND THE PERIODIC TABLE

Of the seven micronutrients, five are transition metals. Four of these (Fe, Mn, Cu, Zn) are cationic and fall in the first transition series of Period 4.

The fifth (Mo) is a heavy metal of the second transition series of Period 5. All these five micronutrients form stable complexes with organic ligands. Complexed to proteins, they function as biological catalysts (metalloenzymes). Except for zinc, they all exist in more than one oxidation state, which enables them to participate in redox reactions and in electron transport. Cobalt and nickel, required for certain biochemical processes in plants, fill the gap between iron and copper in the first transition series.

Boron and chlorine are different from the rest of the micronutrients in their place in the Periodic Table and their roles in plants. Boron is a group III, A metalloid and chlorine, a group VII halogen. Their electronic states do not allow participation in redox reactions or electron transport. Instead, they are involved in structural and regulatory roles.

1.3 GENERAL FUNCTIONS OF MICRONUTRIENTS

Micronutrients play important roles as constituents of organic structures, constituents or activators of enzymes, electron carriers or in osmoregulation. They also function in the regulation of metabolism, reproduction and protection against abiotic and biotic stresses. Roles and deficiency responses of individual micronutrients have been discussed in the following chapters. This chapter gives an overview of their diverse functions.

1.3.1 Micronutrients as Constituents of Organic Structures

The structural role of micronutrients is best exemplified by the role of boron as an essential constituent of higher plant cell walls (O'Neill et al. 2004). Boron is essential for covalent linking of the rhamnogalacturonan II dimers to provide the cell wall with a structure suited to its development (Fleischer et al. 1999) and extension growth (O'Neill et al. 2001). It is also involved in the development of the cytoskeleton (Yu et al. 2003). Boron and zinc are also important to the structure and function of membranes. They are necessary to maintain the structural integrity of the plasmalemma. Inadequacy of either of the two micronutrients leads to alterations in membrane fluidity, causing a marked enhancement in membrane permeability, which manifests itself as an increase in the net efflux of solutes. Boron's effect on membrane permeability is essentially attributed to its role in binding to polyhydroxy groups of the membrane constituents, stabilizing the membrane structures. Recently, involvement of boron in the membrane structure and function has been attributed to its possible role in formation, stability and functioning of membrane rafts (Brown et al. 2002). Zinc's effect on membrane structure and function could be indirect, involving a more basic role in preventing the generation and effective detoxification of reactive oxygen species (Cakmak, 2000). Enhanced accumulation of superoxide radicals (O_2^-) and their reactions with H_2O_2

Group

Period	IA	IIA	IIIB	IVB	VB	VIB	VIIB	VIII	VIII	VIII	IB	IIB	IIIA	IVA	VA	VIA	VIIA	VIIIA
1	1 H 1.008																	2 He 4.003
2	3 Li 6.941	4 Be 9.012											5 B 10.81	6 C 12.01	7 N 14.01	8 O 16.00	9 F 19.00	10 Ne 20.18
3	11 Na 22.99	12 Mg 24.31											13 Al 26.98	14 Si 28.09	15 P 30.97	16 S 32.07	17 Cl 35.45	18 Ar 39.95
4	19 K 39.10	20 Ca 40.08	21 Sc 44.96	22 Ti 47.88	23 V 50.94	24 Cr 52.00	25 Mn 54.94	26 Fe 55.85	27 Co 58.47	28 Ni 58.69	29 Cu 63.55	30 Zn 65.39	31 Ga 69.72	32 Ge 72.59	33 As 74.92	34 Se 78.96	35 Br 79.90	36 Kr 83.80
5	37 Rb 85.47	38 Sr 87.62	39 Y 88.91	40 Zr 91.22	41 Nb 92.91	42 Mo 95.94	43 Tc (98)	44 Ru 101.1	45 Rh 102.9	46 Pd 106.4	47 Ag 107.9	48 Cd 112.4	49 In 114.8	50 Sn 118.7	51 Sb 121.8	52 Te 127.6	53 I 126.9	54 Xe 131.3
6	55 Cs 132.9	56 Ba 137.3	57 La* 138.9	72 Hf 178.5	73 Ta 180.9	74 W 183.9	75 Re 186.2	76 Os 190.2	77 Ir 190.2	78 Pt 195.1	79 Au 197.0	80 Hg 200.5	81 Tl 204.4	82 Pb 207.2	83 Bi 209.0	84 Po (210)	85 At (210)	86 Rn (222)
7	87 Fr (223)	88 Ra (226)	89 Ac~ (227)	104 Rf (257)	105 Db (260)	106 Sg (263)	107 Bh (262)	108 Hs (265)	109 Mt (266)	110 — 0	111 — 0	112 — 0		114 — 0		116		118 — 0

Lanthanide Series*	58 Ce 140.1	59 Pr 140.9	60 Nd 144.2	61 Pm (147)	62 Sm 150.4	63 Eu 152.0	64 Gd 157.3	65 Tb 158.9	66 Dy 162.5	67 Ho 164.9	68 Er 167.3	69 Tm 168.9	70 Yb 173.0	71 Lu 175.0
Actinide Series~	90 Th 232.0	91 Pa (231)	92 U (238)	93 Np (237)	94 Pu (242)	95 Am (243)	96 Cm (247)	97 Bk (247)	98 Cf (249)	99 Es (254)	100 Fm (253)	101 Md (256)	102 No (254)	103 Lr (257)

Fig. 1.1. The Periodic Table of elements, with micronutrient elements shaded in grey

(Haber-Weiss reaction) leads to production of the more toxic oxygen species (OH⁻) which cause damage to the cellular membranes, enzymes and DNA.

1.3.2 Enzyme Action

Perhaps the best known role of micronutrients is their catalytic role, driving and regulating important enzyme-catalyzed reactions. They are known to be a constituent or an activator of several different classes of enzymes but, more importantly, oxidoreductases.

Each, iron, manganese, copper, zinc and molybdenum is a constituent of several oxidoreductases involved in basic metabolic reactions that range from ion uptake, mineral metabolism, photosynthesis, respiration, nitrogen fixation, biosynthetic pathways and stress-defense mechanisms. Manganese and zinc are involved in activities of hydrolases and iron, manganese and zinc in activities of lyases. Manganese is also involved in the activity of glutamine synthetase (a ligase). The reactions catalyzed by these enzymes and their metabolic implications are discussed in chapters 2 to 6.

Micronutrients essentially perform catalytic functions in one of the following ways:

(a) Acquiring high catalytic efficiency by binding to specific proteins, as in the case of copper oxidases and haem enzymes.

(b) Affecting the formation of enzyme-substrate complexes by changing the net charge on the protein or removal of an inhibitory substance. For example, the ratio of zinc and magnesium alters the optima of phosphatase activity (acid or alkaline phosphatases). Iron, zinc and cobalt counteract the cysteine inhibition of yeast aldolase.

(c) Acting as a bridge between the substrate and the enzyme to form an active intermediate (metal-substrate-enzyme) complex. For example, manganese acts as a ligand between amino and carboxyl groups of the substrate and the protein in prolidase.

(d) Functioning as an activator of enzymes, for example, manganese activation of decarboxylases and hydrolases.

Micronutrients may also influence enzyme activities by acting as allosteric effectors.

1.3.3 Charge Carriers

As metal cofactors of enzymes-catalyzing redox reactions (Table 1.2), and metalloproteins functioning as electron carriers, micronutrients participate in the electron transport chains in mitochondria and chloroplasts. They mediate the flow of electrons from one molecule to another. At certain steps, the transport of electrons is coupled to generation of energy-rich molecules (ATP) and production of reducing power (NAD(P)H), needed to drive key metabolic reactions. The different oxidation states of manganese enables charge accumulation to produce the reducing power required for

Table 1.2 Higher plants enzymes having micronutrient cofactors

Enzyme (EC)	Micronutrient cofactor/activator
Oxidoreductases:	
Superoxide dismutases (EC 1.15.1.1)	Cu-Zn, Mn, Fe
Lipoxygenase (EC 1.13.11.12)	Fe
Alternative oxidase	Fe
Nitrite reductase (EC 1.7.7.1)	Fe
Fdx-Thioredoxin reductase	Fe
Sulphite reductase (EC 1.8.3.1)	Fe
Succinate dehydrogenase (EC 1.3.99.1)	Fe
NADH-Q oxidoreductase (EC 1.6.99.5) (NADH dehydrogenase)	Fe
Catalase (1.11.1.6)	Fe
Peroxidases (1.11.1.7)	Fe
Cytochrome c oxidase (EC 1.9.3.1)	Fe,Cu
Cytochrome c reductase (EC 1.9.99.1) (Q cytochrome c oxidoreductase, Cyt bc_1)	Fe
Glutamate synthase (EC 1.4.7.1, 1.4.1.14)	Fe
NAD^+ Malic enzyme (EC 1.1.1.39)	Mn
$NADP^+$ Malate dehydrogenase (EC 1.1.1.40)	Mg, Mn
Isocitrate dehydrogenase (EC 1.1.1.41, 1.1.1.42)	Mg, Mn
Catechol oxidase (EC 1.10.3.1)	Cu
Laccase (EC 1.10.3.2)	Cu
Ascorbate oxidase (EC 1.10.3.3)	Cu
Monophenol monoamine oxidase (EC 1.14.18.1)	Cu
Copper amine oxidase (EC 1.4.3.6)	Cu
Alcohol dehydrogenase (EC 1.1.1.2)	Zn
Glutamine dehydrogenase (EC 1.4.1.3)	Zn
Nitrate reductase (EC 1.6.6.1)	Mo
Xanthine dehydrogenase (EC 1.1.1.204)	Mo
Aldehyde oxidase (EC 1.2.3.1)	Mo
Sulphite oxidase (EC 1.8.3.1)	Mo
Hydrolases	
Allantoate amidohydrolase (EC 3.5.3.9)	Mn
Carboxypeptidase A (EC 3.4.17.1)	Zn
Lyases	
Aconitase (EC 4.2.1.3)	Fe
Enolase (EC 4.2.1.11)	Mg, Mn
Phospho*enol*pyruvate carboxylase (EC 4.1.1.31)	Mg, Mn
Carbonic anhydrase (EC 4.2.1.1)	Zn
Ligases	
Glutamine synthetase (EC 6.3.1.2)	Mg, Mn

photooxidation of water in the initial reaction of PS II. Likewise, the different oxidation states of molybdenum enable it to catalyze the enzymatic reduction of N=N bond in N_2 to NH_4.

Mitochondrial Electron Transport

Iron and copper function as cofactors of the electron transport proteins of the inner mitochondrial membrane (Fig 1.2). NAD^+-isocitric dehydrogenase activated by Mg^{2+} or Mn^{2+} catalyzes the reduction of NAD^+ to NADH in the mitochondrial matrix.

Photosynthetic Electron Transport

As a constituent of oxygen-evolving complex of PS II, manganese is involved in the oxidation of water. Iron and copper are involved in electron

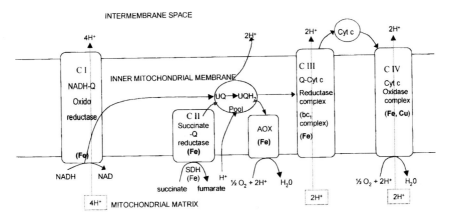

Fig. 1.2. Mitochondrial electron transport chain, showing micronutrients (Fe, Cu) as components of electron transport carriers

transport through photosystem II and photosystem I and in coordinating the flow of electrons from PS II to PS I.

The non-haem protein ferredoxin accepts the electrons from P 700 and donates them to ferredoxin-NADP reductase. This enzyme catalyzes the production of NADPH, which serves as a reductant in Calvin cycle

1.3.4 Osmoregulation

Plant nutrient status of micronutrients is known to influence plant water relations. Chlorine plays a direct role in osmoregulation. It is involved in cell osmotic relations and turgor-dependent extension growth of cells. Chlorine and potassium fluxes across leaf guard cell plasma membrane induce turgor changes that regulate stomatal opening. Chlorine also plays an important osmoregulatory role under conditions of salt stress. The

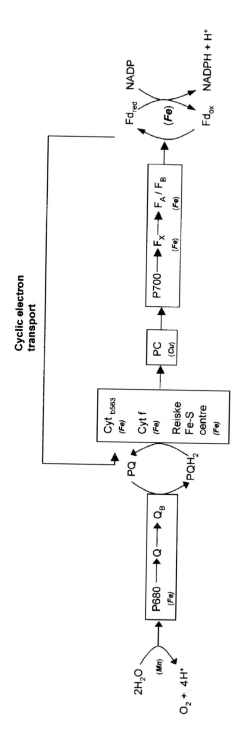

Fig. 1.3. Scheme of photosynthetic electron transport showing micronutrient components

osmoregulatory concentrations of chlorine are, however, of a much higher magnitude than the concentrations at which other micronutrients perform their critical roles.

1.3.5. Secondary Metabolism, Growth Hormones and Signalling Molecules

An important role of micronutrients involves their functioning in biosynthetic pathways. Many enzymes containing micronutrient cofactors catalyze reactions in biosynthesis of secondary metabolites. Precursors for the synthesis of aromatic amino acids are synthesized via shikimate pathway, whose initial reaction involves the condensation of phosphoenolpyruvate and erythrose-4-phosphate to provide 3-deoxy-D-arabinoheptulosonolate-7 phosphate (DAHP), catalyzed by the manganese-activated enzyme DAHP synthase (Herrmann, 1995; Hermann and Weaver, 1999). Production of the lignin precursor p-coumaric acid from trans-cinnamic acid is catalyzed by NADPH-Cytochrome P450 dependent haem enzyme, cinnamate-4-hydroxylase. Zinc-activated cinnamyl dehydrogenase catalyzes the conversion of p-coumaraldehyde and other cinnamic aldehydes to cinnamic alcohol (monolignols), which are subsequently polymerized to lignin.

Biosynthesis of gibberellins, lately recognized as major signalling molecules in plants (Sun and Gubler, 2004), involves the activity of several enzymes activated by micronutrients. Synthesis of the gibberellin precursor *ent*-kaurene involves catalysis by kaurene synthase, which is activated by Mn^{2+}, Co^{2+} or Mg^{2+}. The next three steps involving conversion of *ent*-kaurene to *ent*-kaurenoic acid are catalyzed by *ent*-kaurene oxidase, which is a P450 haem monooxygenase (Chapple, 1998). Several 2-oxodependent dioxygenases catalyze the conversion of GA_{12} aldehyde to gibberellins and the interconversions of the latter, including those leading to production of their bioactive forms (Hedden and Phillips, 2000; Schomburg et al. 2002) (Fig. 1.4).

Three enzymes of jasmonic acid biosynthesis contain iron cofactors. Lipoxygenase is a non-haem iron-containing dioxygenase (Feussner and Wasternack, 2002); fatty acid hydroperoxide lyase and allene oxide synthase are cytochrome P450 dependent monooxygenases (Chapple, 1998). Terminal steps in the biosynthesis of ethylene and abscissic acid are also catalyzed by enzymes having micronutrient cofactors. 1-aminocyclopropane-1-carboxylate oxidase, which catalyzes the synthesis of ethylene, is specifically activated by iron (Prescott and John, 1996). Aldehyde oxidase, which catalyzes the synthesis of abscissic acid from abscissic aldehyde, is a molybdoprotein (Romao et al. 1995). Synthesis of flavones and anthocyanin also involves catalysis by cytochrome P450 haem monooxygenases (Chapple, 1998).

12

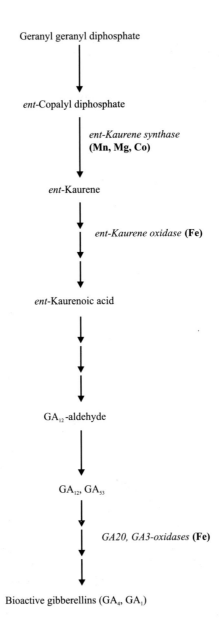

Geranyl geranyl diphosphate

↓

ent-Copalyl diphosphate

ent-Kaurene synthase
(Mn, Mg, Co)

ent-Kaurene

ent-Kaurene oxidase **(Fe)**

ent-Kaurenoic acid

GA_{12}-aldehyde

GA_{12}, GA_{53}

GA20, GA3-oxidases **(Fe)**

Bioactive gibberellins (GA_4, GA_1)

Fig. 1.4. Steps in gibberellin biosynthesis pathway catalyzed by micronutrient requiring enzymes

1.3.6 Protective Role

Micronutrients are known to influence the generation and detoxification of reactive oxygen species that induce oxidative stress and play an important role in signal transduction (Apel and Hirt, 2004). Enzymes having micronutrient cofactors function as a part of the antioxidant system, offering protection against damage from excessive generation of reactive oxygen species (ROS) during electron transport in mitochondria and chloroplasts (Fridovich, 1986; Elstner, 1991; Asada, 1994). Superoxide dismutases, which have micronutrient cofactors Fe-SOD, Mn-SOD, Cu-Zn SOD, constitute the first line of defense against the ROS (Alscher et al. 2002). Mitochondria have high activities of Mn-SOD and chloroplasts show high activities of Cu-Zn and Fe-SOD. The cytosol and the apoplasm also show Cu-Zn SOD activity. Thus, the superoxide ions (O_2^-) are effectively converted to H_2O_2 within close vicinity of the site of their generation. Accumulation of H_2O_2 in toxic concentrations is prevented by its reduction to water in reactions catalyzed by the haem enzymes ascorbate peroxidase (APX) and catalase (CAT). APX, which uses ascorbate as its specific electron donor, reduces H_2O_2 to water with concomitant production of monodehydroascorbate, which is spontaneously disproportionated to dehydroascorbate. Working in combination with the ascorbate–glutathione cycle, APX functions as a potent scavenger of H_2O_2 in chloroplasts (Asada, 1992, 1997). Catalase, which carries out rapid breakdown of H_2O_2, is highly concentrated in peroxisomes, wherein conversion of glycolate to glyoxylate also adds to accumulation of H_2O_2. Thus, acting as cofactors of antioxidant enzymes SOD, APX and CAT, the micronutrients (Cu, Zn, Fe, Mn) contribute to defense against oxidative stress. Overexpression of superoxide dismutases in transgenics has been shown to enhance their tolerance to oxidative stress (Gupta et al. 1993; Perl et al. 1993; Slooten et al. 1995; Van Camp et al. 1996).

A decrease in the activity of the antioxidant enzymes due to limited availability of the metal cofactors weakens the antioxidant defense mechanism and exposes the plants to greater damage from the ROS. Cakmak (2000) has provided a comprehensive update on the role of zinc in protection of plants against oxidative stress.

Another example of the protective role of micronutrients is the role of iron as a cofactor of choline monooxygenase (CMO) in glycine betaine accumulating plants. CMO catalyzes the first step in the biosynthesis of glycine betaine which acts as osmoprotectant, stabilizing the quarternary structure of proteins and maintaining the structural integrity of cellular membranes under stressful conditions imposed by high salinity and temperature (Graham, 1995).

14

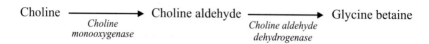

Choline   ⟶   Choline aldehyde   ⟶   Glycine betaine
       Choline   *Choline aldehyde*
    monooxygenase   *dehydrogenase*

Not only are micronutrients involved in providing protection against abiotic stresses, they also provide protection against pathogenic infections (Graham, 1983; Graham and Webb, 1991; Huber and Graham, 1999). This, micronutrients may do in two ways. Manganese, copper and boron are involved in the metabolism of phenolic compounds and their polymerization to lignins and suberins, which strengthens the plant cell walls, making them less susceptible to penetration by pathogens (Mehdy, 1994). Another way that the micronutrients offer protection against pathogenic infections is through enhanced production of H_2O_2, which functions as a cytotoxic compound against the pathogens and also activates the defense mechanism of the host (Lamb and Dixon, 1997; Orozeo-Cardinas et al. 2001). Copper, as a constituent of diamine oxidases, plays an important role in generation and regulation of H_2O_2 concentration in response to attack by pathogens (Fedrico and Angelina, 1991, Rea et al. 1998).

1.3.7 Regulatory Role

Functions of micronutrients are not localized to a particular plant part, cell or cell compartment. Therefore, they have to be available at all the functional sites in concentrations commensurate with their biochemical requirement. This necessitates a mechanism for their homeostasis. This may involve induction of high-affinity transport systems in response to deficiency and a mechanism for their subcellular sequestration under conditions of overload. High-affinity uptake mechanisms for uptake of iron are upregulated on sensing iron deficiency (Marschner and Romheld, 1994). The level of iron supply may also determine the tissue and cellular mobilization of iron (Bereczky et al. 2003, Thomine et al. 2003). The active process of boron uptake is induced in response to boron deficiency (Dannel et al. 2002). Zinc regulates the uptake of phosphorus by regulating the genes encoding high-affinity phosphate transport (Huang et al. 2000). Several TF III A-type zinc finger proteins such as SUPERMAN, AtZFP1, PetSPL3 and Bc ZFP1 are implicated in the developmental regulation of various floral and vegetative organs (Takatsuji, 1999). Zinc is also reported to be involved in the prevention of cell death in plants (Dietrich et al. 1997). In *Arabidopsis*, a zinc finger protein functions as a negative regulator in plant cell death.

1.3.8 Role in Reproduction

So far, the predominant sporophytic generation of higher plants has been the main focus of investigations on the roles and deficiency responses of micronutrients. It is being increasingly realized that poor reproductive yield of plants subjected to deficiency of micronutrients cannot be entirely attributed to their effect on photosynthetic efficiency and/or partitioning of assimilates. Reproductive development of plants may be severely limited even when deficiency is induced, nearing onset of reproductive phase, by which time plants have accumulated enough photosynthates. Many studies on the reproductive development following deprivation of micronutrient supply to plants point to a more direct involvement of micronutrients in reproductive development (Graham, 1975; Dell, 1981; Sharma et al. 1990; Sharma et al. 1991; Pandey et al. 2006). Development of different floral organs in *Arabidopsis*, petunia and Chinese cabbage has been suggested to involves TF III A type zinc finger control of cell division and/or expansion of particular cell types (Takatsuji, 1999) There are also reports suggesting that the role of zinc finger polycomb group proteins in development of female gametophyte and seed development (Grossniklaus et al. 1998; Brive et al. 2001). Reproductive development may also be inhibited because of poor delivery of micronutrients in the reproductive tissues due to some morphological limitations or poor transport (Brown et al. 2002; Takahashi et al. 2003). Studies using transgenic tobacco plants (*naat* transgenics) and crosses between the *naat* transgenics and the wild type (Takahashi et al. 2003) show severe limitation in delivery of Zn, Cu and Mn into the reproductive parts due to lack of nicotianamine, which has a role in chelation and long-distance transport of the cationic micronutrients (Welch, 1995; Von Wiren et al. 1999).

There is a quantitative requirement of the micronutrients for optimal performance of their roles. Deviations from the optimum are reflected in aberrations in structure, function, and developmental and adaptive responses of plants. These issues are addressed in the following chapters of Part I.

CHAPTER 2

IRON

2.1 GENERAL

Iron is the third most abundant element in the earth's crust. It exists in two oxidation states—ferric (Fe^{3+}) and ferrous (Fe^{2+}). The easy inter-convertibility of the two oxidation states ($Fe^{3+} \leftrightarrow Fe^{2+}$) confers on the metal the ability to participate in both oxidation-reductions and electron transport. Iron acquires high biological activity on binding to proteins. Iron proteins play important roles as donors or acceptors of electrons. Bound in different forms, iron forms the cofactor or prosthetic group of several enzymes. Many iron enzymes, recently cloned using molecular biology techniques, are known to play a role in metabolism of fatty acids and biosynthesis of terpenoids, growth hormones and signalling molecules. Iron is involved both in generation of active oxygen species and their detoxification. Iron homeostasis is critical to sustain metabolism. Redox changes during electron transport involving iron proteins and iron involvement in biosynthesis of jasmonic acid and ethylene are suggestive of a role of iron in signal transduction.

Iron status of plants is critical to its uptake and compartmentalization. There is growing evidence to show that iron plays a homeostatic role to ward against abiotic stresses. Cloning of genes encoding enzymes of the antioxidant system (Fe-SOD, APX) and their overexpression in transgenics has been shown to enhance tolerance to oxidative stress (Slater et al. 2003). Overexpression of choline monooxygenase, a cytochrome P450 monooxygenase, offers osmoprotection against salt and temperature stress.

2.2. UPTAKE, TRANSPORT, DISTRIBUTION

Plants show a taxonomic distinction in the way they mobilize iron from the rhizosphere and transport it across the plasmalemma (Brown and Jolley, 1989; Romheld and Marschner 1986a,b; Marschner et al. 1986a). Iron deficiency induces two distinct and mutually exclusive mechanisms or strategies for iron acquisition (Bienfait, 1988; Marschner and Romheld, 1994); one by the dicotyledons and monocotyledons except the graminaceae (Strategy I); and the other by the members of the graminaceae (Strategy II). Strategy I plants use a reduction-based uptake mechanism. Roots of plants belonging to this strategy reduce external Fe^{3+} to Fe^{2+} and transport iron across the plasmalemma only in the reduced form (Fe^{2+}). Strategy II plants lack this mechanism or express it at a very low level (Zacharieva and Romheld, 2001). They mobilize iron from the rhizosphere by producing and releasing ferric (Fe^{3+}) solubilizing compounds termed phytosiderophores (Takagi, 1976; Romheld and Marschner, 1986). The phytosiderophores, exemplified by mugineic acids, are non-proteinogenic amino acids produced in response to iron deficiency. Strategy II plants take up iron in the form of Fe^{3+} chelates of mugineic acids.

Strategy 1 Plants

Strategy I plants, adapted to reduction-based uptake of iron, respond to iron deficiency stress by inducing increased reduction of soluble Fe^{3+} to Fe^{2+}. Several morphological and physiological changes in roots have been suggested to contribute to the increased reduction of iron. The more important of these are:

(a) Changes in root morphology, such as formation of transfer cells (Kramer et al. 1980; Romheld and Marschner, 1981b; Landsberg, 1982, 1986; Romheld and Kramer, 1983).
(b) Exudation of organic acids (Tiffin, 1966).
(c) Enhanced secretion of reducing substances (Brown, 1978; Hether et al. 1984).
(d) Enhanced proton extrusion (Brown, 1978; Landsberg, 1989; Alcantara et al. 1991; Rabotti and Zocchi, 1994; Rabotti et al. 1995).
(e) Reduction of Fe^{3+} to Fe^{2+} at the plasmalemma (Chaney et al. 1972; Bienfait et al. 1983; Brüggemann et al. 1990; Alcantara et al. 1991).

Bienfait (1985) first showed that reduction of extracellular Fe (III) chelate to Fe (II) chelate is catalyzed by the Fe (III) chelate reductase or a 'turbo' reductase, induced in the plasmalemma of root epidermal cells in response to iron deficiency stress. Reduction of iron by Fe^{3+} chelate reductase has since been established as the first step (Marschner and Romheld, 1994) and the prime factor essential for uptake of iron by strategy I plants. (Moog and Bruggemann, 1995; Yi and Guerinot, 1996). Yi and Guerinot (1996) presented genetic evidence to show that induction of root Fe (III) chelate reductase activity is essential for iron uptake under conditions of iron deficiency. *Arabidopsis* mutants *frd-1, frd-2* and *frd-3* (*Ferric Reductase Defective 1, 2* and *3*) that lack the ability to induce Fe^{3+} chelate reductase activity show up the defect in uptake of iron (Yi and Guerinot, 1996). Similarity between iron uptake mechanisms in yeast, which are fairly well understood, and higher plants paved the way for characterization of several putative iron transporters in plants. Taking the advantage of sequence similarity to yeast Fe^{3+} chelate reductases FRE1 and FRE2, Robinson et al. (1999) made the first characterization of Fe^{3+} chelate reductase gene from *Arabidopsis*. This was named *FRO2* (*Ferric Reductase Oxidase 2*) and, on the basis of sequence similarity with the human phagocyte NADPH gp 91-phox oxidoreductase, identified as a member of the flavocytochrome superfamily. *FRO2* encodes the integral membrane Fe^{3+} chelate reductase, which transfers electrons from the cytosolic NADPH to the enzyme bound FAD through the haem groups to Fe^{3+} on the opposite (extracellular) side of plasmalemma resulting in reduction of Fe^{3+} to Fe^{2+}. It has been shown that *At FRO2* is expressed in roots and its mRNA levels are upregulated by iron deficiency (Robinson et al. 1999). Loss of function mutations of *FRO2* show a decrease in Fe^{3+} chelate reductase activity, produce chlorosis and inhibit plant growth under iron deficiency; *FRO2* expression in iron-uptake defective *Arabidopsis* mutant *frd-1-1* restores Fe^{3+} chelate reductase activity. Another Fe^{3+} chelate reductase gene *FRO1* has been cloned from roots of iron deficient pea (cv. Sparkle) (Waters et al. 2002). This gene *Ps FRO1* shows close sequence similarity to *At FRO2* and its mRNA levels are correlated with Fe^{3+} chelate reductase activity. Even though most abundant in roots, more so in the outer epidermal cells, *Ps FRO1* is also expressed in leaves and root nodules suggesting its possible role in whole plant distribution of iron. (Waters et al. 2002).

The contributions of the morphological and physiological changes associated with enhanced reduction of iron in strategy I plants to iron uptake have been a subject of discussion (Jolley et al. 1996; Ma and Nomoto, 1996; Pearson and Rengel, 1997; Rengel, 1999; Schmidt, 1999; Curie and Briat, 2003). In a recent study, Zheng et al. (2003) reported induction of root Fe (III) chelate reductase in iron efficient red clover (*Trifolium pretense* L) cultivar Keenland within 24 h of exposure to iron deficiency, within which period the roots did not show any enhancement in H^+ extrusion. This suggested that proton extrusion is not a pre-condition for reduction of iron. It is, however, certain that the external pH plays an important role in solubilizing the sparingly soluble Fe^{3+} in the rhizosphere. Maximum reduction of Fe^{3+} occurs around pH 5.0. High pH of the external solution, including the apoplasmic fluid, inhibits Fe^{3+} reduction (Toulon et al. 1992; Susin et al. 1999; Manthey et al. 1996). Some iron deficiency responses are possibly regulated by hormones such as ethylene. Romera et al. (1999) reported an increase in ethylene production in response to iron deficiency and showed that it induced changes in root morphology and Fe (III) reductase activity, contributing to enhanced uptake of iron. Schmidt et al. (2000) observed enhanced production of ethylene and alterations in root morphology of iron-deficient plants of *Arabidopsis*, which favoured enhanced uptake of iron, but these changes had little effect on Fe (III) reductase activity.

Recently, Zaid et al. (2003) have also reported ethylene-induced changes in root morphology, such as the formation of cluster roots, that contribute to increased uptake of iron in response to iron deficiency.

Transport of iron, reduced at the external surface of the plasmalemma of strategy I plants, into the cytosol involves a high-affinity transport system. In the recent past, concerted efforts have been made to identify and characterize possible plasmalemma-bound Fe^{2+} transport proteins. Making use of advancements in molecular biology techniques and functional complementation of iron-uptake deficient yeast mutants, several transporter genes, belonging to different gene families, have been identified. The first iron regulated transporter IRT1 (Iron Regulator Transporter 1), belonging to the Z1P (ZRT – 1RT like Proteins) family was identified from *Arabidopsis* (Eide et al. 1996). The At IRT1 was shown to complement the iron uptake deficiency yeast double mutant *fet3 fet4*. Iron deficiency has been shown to cause accumulation of At IRT1 transcript and recovery from this deficiency leads to a decline in the level of the transcript (Connolly et al. 2002). Vert et al. (2002) have shown that IRT1 mutant line of *Arabidopsis* is not viable unless supplemented with high iron supply. The knock-out (T-insertion) mutant of At IRT1 *(irt1)* develops foliar symptoms of iron deficiency and developmental defects in different plant parts (Henriques et al. 2002; Vert et al. 2002). Grotz and Guerinot (2000) have isolated another putative iron transporter gene, *IRT2* from *Arabidopsis* and showed that it is expressed in root epidermal cells in response to iron deficiency. Eckhardt et al. (2001) have characterized two cDNAs from a library constructed from roots of iron deficient tomato plants and designated these *Le IRT1* and *Le IRT2*. It has been shown that both *Le IRT1* and *Le IRT2* complement iron-uptake mutants of yeast and are predominantly expressed in roots; but only *Le IRT1* is upregulated under iron deficiency. An *IRT* ortholog (*RIT1*) isolated from pea seeds has been shown to encode a protein that is 63% identical to *At IRT1* and is upregulated by iron deficiency (Cohen et al. 1998). IRT1 is now considered as the main transporter involved in high-affinity iron uptake by roots of dicotyledonous (Strategy I) plants under conditions of iron deficiency (Connolly et al. 2002; Vert et al. 2002).

Some Nramp family transporters have also been suggested to be involved in uptake of iron (Belauchi et al. 1997; Curie et al. 2000; Thomine et al. 2000). It has been shown that *Arabidopsis Nramps 1, 3* and *4* complement iron-transport defect in yeast mutants and are upregulated under iron deficiency (Thomine et al. 2000). *Nramp 3* and *Nramp 1* have recently been shown to be involved in vascular transport of iron and other co-substrates (Thomine et al. 2003; Bereczky et al. 2003). Thomine et al. (2003) showed that the *Nramp 3* gene is expressed in vascular bundles of roots, stems and leaves and further suggested that the protein encoded by it plays a role in long-distance transport of iron and other co-substrates (manganese and cadmium). Bereczky et al. (2003) cloned a *Nramp 1* gene from tomato *(Le Nramp 1)* and showed it to be specifically expressed in roots and upregulated in response to iron deficiency. Bereczky et al. (2003) showed the Le Nramp1 to be localized in the vesicular parenchyma of the root of iron deficient plants and suggested that the transporter is possibly involved in the mobilization of iron in vascular tissues of plants.

While it is unequivocally accepted that the strategy 1 response is activated by limitation in iron availability, it is debatable as to whether the deficiency of iron is sensed and transformed into a signal-regulating iron uptake by the root or the shoot (Curie and Briat, 2003). Early experiments involving reciprocal grafting of the chlorotic tomato mutant *T 3238 fer* lacking in the ability to activate the strategy I response to the wild type showed that the *fer* gene evoking the iron-deficiency response is required in root and not in shoot (Brown et al. 1971; Brown and Ambler, 1973). Studies with pea mutant *brz* also supported this (Kneen et al. 1990). In consonance with these findings, Bienfait et al. (1987) reported that irrespective of whether potato tubers were sprouted or not, their roots showed the same iron-deficiency response. Ling et al. (1996, 2002) provided genetic and molecular evidences for the root being the organ involved in perceiving deficiency of iron and activating the deficiency response. Ling et al. (1996) showed that the *fer* gene was required exclusively in the root. Later, they isolated the *fer* gene by map-bound cloning from roots of iron-deficient tomato plants and demonstrated that it encodes a regulatory protein (transcription factor) containing a bHLH domain. When the *fer* gene mutant, lacking in ability to activate the iron-deficiency response, was complemented with the *fer* gene, the ability to activate the iron-deficiency responses such as upregulation of Fe^3 chelate reductase and induction of *Le IRT1* expression, were restored. Ling et al. (2002) suggested that the *fer* gene is expressed in a cell-specific pattern at the root tip, where it senses the availability of iron and activates the iron-deficiency response through transcriptional control. Rogers and Guerinot (2002) have characterized another gene *FRD3* from roots of iron-deficient *Arabidopsis*. The *FRD3* gene belongs to the multidrug and toxin efflux (MATE) family and is detectably expressed in the roots only. It has been shown that FRD3 functions in regulation of iron uptake and homeostasis and that the *FRD 3* mutants are defective in iron deficiency signalling and iron distribution (Rogers and Guerinot, 2002). In another recent study, Bereczky et al. (2003) have demonstrated that the tomato iron transporter genes *Le nramp 1* and *LeIRT 1* are expressed specifically in the roots and that both are downregulated in roots of the *fer* mutant.

Contrary to the above information, several workers (Maas et al. 1988; Romera et al. 1992; Grusak, 1995, Grusak and Pezeshgi, 1996) have presented evidence for a role of shoot in activating the iron-deficiency response. Romera et al. (1992) showed that exposure of half of the roots to an iron-deficient nutrient solution in a split-root experiment induced iron-deficiency response by the other half of the roots, suggesting signal transport through the shoot. Reciprocal grafting experiments by Grusak and Pezeshgi (1996) showed that shoot, not the root, determined the phenotypic response of the iron-accumulating pea mutants *brz* and *dgl*. These mutants possess a constitutive Fe^{3+} chelate reductase in roots, which can be downregulated under iron sufficient condition in response to a shoot-derived signal.

Waters et al. (2002) have shown that the pea Fe^{3+} chelate reductase gene *FRO1* is differentially regulated in root and shoot of the wild type and the iron-accumulating mutants *brz* and *dgl*. In both the wild type and the mutants, *FRO1* expression and Fe^{3+} chelate reductase activity in shoot exhibit the iron-deficiency response. It is the same in the roots of the wild type (Sparkle), but in the iron-accumulating mutants *brz* and *dgl*, both *FRO1* expression and Fe^{3+} chelate reductase activity are constitutive. These findings are in accord with the results of the reciprocal grafting experiments of Grusak and Pezeshgi (1996). The expression of *FRO1* in roots and shoots is affected by different signals or iron-sensing mechanisms. The *FRO1* expression in roots of the wild type is regulated by a shoot-derived signal generated by the shoot iron concentration (Waters et al. 2002).

Recent findings of Zheng et al (2003) show temporal differences in the root- and shoot-generated signals regulating the iron-deficiency response. Deficiency of iron is first perceived by the root, which activates rapid (within 24 h) induction of Fe^{3+} chelate reductase activity associated with proton extrusion. Subsequently, the Fe^{3+} chelate reductase activity is upregulated in response to a shoot-derived signal, possibly transmitted from the shoot to the root through the iron concentration in the phloem, as was conceived earlier by Maas et al. (1988). There is also evidence to show the involvement of separate signals for activation of different iron-deficiency responses. Schikora and Schmidt (2001) showed that in pea mutants *brz* and *dgl*, Fe^{3+} chelate reductase activity is high regardless of iron supply but transfer cell formation is induced in response to iron deficiency, which suggests that separate signals may be required for induction of Fe^{3+} chelate reductase activity and transfer cell formation.

Strategy II Plants

Strategy II plants take up iron as ferric complexes of the mugineic acid family phytosiderophores (Fe^{3+}-PS). The first step in the uptake of iron involves the synthesis of mugineic acids capable of chelating Fe^{3+}- (Mori et al. 1990; Ma and Nomoto, 1996). On the basis of short-term experiments with iron-deficient barley and various chelating substances, including different metals and mugineic acids (MA), Ma et al. (1993) demonstrated that formation of MA-Fe^{3+}- complex is a prerequisite for its recognition by the specific Fe^{3+}- transport system in barley roots. The mugineic acids have been shown to chelate Fe^{3+} with their amino and carboxyl groups (Ma and Nomoto, 1996).

Mugineic acids are synthesized from methionine (Mori and Nishizawa, 1987; Shojima et al. 1990; Ma et al. 1995) (Fig. 2.1). The first two reactions of the MA biosynthesis pathway involve the synthesis of NA from Met and are ubiquitous, in no way confined to strategy II plants. In the first reaction of the pathway, Met is converted to S-adenyl–L–methionine (SAM) by SAM synthase (ATP: methionine S-adenyl transferase). The second reaction involves the isomerization of the three molecules of SAM to one molecule of NA, and is catalyzed by nicotianamine synthase (NA synthase). The genes encoding NA synthase have been cloned and characterized from both graminaceous as well as dicotyledonous plants (Shojima et al. 1990; Higuchi et al. 1994,1999; Herbick et al. 1999; Ling et al. 1999). Further steps for the synthesis of MAs from NA are confined to graminaceous (strategy II) plants and the genes encoding the enzymes catalyzing these reactions have been cloned and characterized from plants adapted to this strategy (Okumura et al. 1994; Higuchi et al. 1999; Takahashi et al. 1999; Kobayashi et al. 2001). In a reaction catalyzed by nicotianamine aminotransferase (NAAT), the graminaceous plants convert NA to an unstable intermediate (NA-3'-oxoacid), which is rapidly reduced to 2-deoxymugineic acid (2'-DMA) involving catalysis by DMA synthase. The barley NAAT is encoded by two genes, *Naat A* and *Naat B* (Takahashi et al. 1999). Introduction of the barley *NAAT* genes into rice enhances the secretion of phytosiderophores and contributes to resistance of the transgenics against chlorosis in alkaline soils (Takahashi et al. 2001). Other mugineic family phytosiderophores, viz., -3-epihydroxy mugineic acid (epi-HMA) and 3-epihydroxy, 2-deoxymugineic acid (epi-HDMA) are synthesized by hydroxylation of DMA. The hydroxylation reaction is catalyzed by 2-oxoglutarate-dependent dioxygenases encoded by two genes *Ids3* and *Ids2*. These genes have since been isolated from the roots of iron- deficient barley (Okumura et al. 1994; Nakanishi et al. 2000; Kabayashi et al. 2001).

All graminaceous species secrete MA-phytosiderophores under conditions of iron deficiency but differ in the secretion of the specific MAs (Rengel, 1999). Wheat, rice, corn and soybean genotypes are known to release only 2-DMA.Barley, oats and rye; on the other hand, they do secrete additional MAs such as 3-hydroxy mugineic acid (HMA) and 3-epihydroxy mugineic acid (epi-HMA). These MAs, being more effective in mobilizing iron than 2-DMA, contribute to greater tolerance of barley and rye to iron chlorosis. It has been suggested that young rice plants are prone to iron chlorosis because they secrete only 2-DMA, and this also ceases early, resulting in iron deficiency and consequential damage to roots (Mori et al. 1991). The release of phytosiderophores by the graminaceous plants under conditions of iron deficiency has been shown to follow a distinct diurnal rhythm (Marschner et al. 1986).

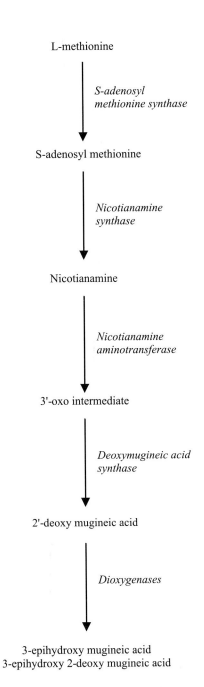

Fig. 2.1. Biosynthetic pathway of mugineic acid phytosiderophores (After Takahashi et al. 2003)

Information about the mechanism involved in secretion of MA-phytosiderophores into the rhizosphere, wherein they solubilize the sparingly soluble Fe^{3+} by chelation, is rather limited. According to Sakaguchi et al. (1999), the MA-phytosiderophores having high affinity for Fe^{3+} are secreted through the plasmalemma via anion channels involving Na^+/MA symport. Blocking of anion channels in barley root plasma membrane has been shown to result in a sharp decline in secretion of MAs. Based on DNA micro-array of gene expression in rice, Negishi et al. (2002) have suggested the possible involvement of vesicles derived from endoplasmic reticulum in secretion of phytosiderophores under conditions of iron- deficiency. Microbial siderophores in the rhizosphere are also reported to contribute to Fe^{3+} mobilization by plants, particularly when iron supply is limiting (Zhang et al. 1991c; Masalha et al. 2000).

Uptake of ferric-phytosiderophore complex (Fe^{3+}-PS), in which form iron is transported across the plasmalemma in graminaceous (Strategy II) plants, also involves a high-affinity transport system (Marschner and Romheld, 1994; Curie and Briat, 2003). Initial information about Fe^{3+}-PS transporters came from a study of the iron-deficiency response of iron-uptake defective maize mutant *ys1*. The iron uptake defect in *ys1* was caused by a defect in the uptake of Fe^{3+}-PS (Von Wiren et al. 1994). This implied that the Fe^{3+}-PS transporter was encoded by *YS1*. Proof of this came from cloning of *YS1* gene (Curie et al. 2001). Molecular characterization of *YS1* showed it to belong to a subclass of oligopeptide transporter (opt) family. It was shown that *YS1* is expressed both in roots and shoots and that its expression is upregulated by iron deficiency. Enhanced accumulation of the *YS1* transcript in shoots led Curie et al. (2001) to propose its role in distribution of iron to the whole plant. Yamaguchi et al. (2002) have identified an iron transporter gene *ID17*, localized to the tonoplast in roots of iron-deficient barley plants. It has been shown that the transcript levels of *ID17* in roots are strongly correlated with iron nutritional status and its expression in roots (only) is upregulated by iron-deficiency. Yamaguchi et al. (2002) have suggested *ID17* involvement in iron transport across the tonoplast.

As a rule, the graminaceous plants lack in reduction-based mechanism for uptake of iron or this is expressed at a very low level (Zacharie and Romheld, 2001), but this may have exceptions. Iron-deficient plants of rice, a typical strategy II plant, has been recently shown to express a functional Fe (II) transporter system, as is characteristic of strategy I plants (Bughio et al. 2002). Bughio et al. (2002) isolated a cDNA clone from roots of iron-deficient rice plants, and showed that its product functions as an Fe^{2+} transporter. The cDNA clone, named *Os IRT1*, showed high homology to *At IRT1*. Its expression in iron-uptake and ferric-reductase defective yeast strain YH003 reversed the growth defect of the strain. It has also been shown that the *Os IRT1* gene is predominantly expressed in roots and induced by deficiency of iron (and copper) (Bughio et al. 2002).

While there is a clear distinction between the induction of mechanisms for increased mobilization and uptake of iron in response to iron deficiency, plants adapted to the two strategies show little distinction in the uptake mechanism when iron supply is sufficiently met (adequate). Under conditions of adequate iron supply, even roots of the strategy II plants (graminaceae) can carry out Fe^{3+} reduction, possibly involving the same reductive system as in strategy I plants (Bagnaresi et al. 1997).

Iron transport from roots to the aerial parts takes place through the xylem, along the transpiration stream. Presence of ferric complexes with organic compounds, particularly citrate, in the xylem exudates suggested that iron is transported as ferric citrate (Tiffin, 1996; Cataldo et al. 1988). This implies that the iron taken up in the reduced form (Fe^{2+}) is reoxidized in the cytoplasm prior to its long-distance transport (Brown and Jolley, 1989). Phloem transport of iron also takes place in an organically bound form. Maas et al. (1988) showed that phloem transport of iron takes place in the form of Fe^3 complexes. The main carrier of iron in the phloem is nicotianamine (Pich et al. 1991; Becker et al. 1992; Stephan and Scholz. 1993; Stephan et al. 1996). Nicotianamine (NA), a non proteinogenic tripeptide derived from methane, chelates both Fe^{3+} and Fe^{2+} but it is only as the Fe^{3+} -NA chelate that iron is transported through the phloem.

Evidence based on physiological characterization of the maize mutant *YS1* suggests that Fe^{3+} - NA transport through the phloem is mediated via a high-affinity transporter expressed in response to perception of iron-deficiency signal (Curie and Briat, 2003, references therein). In addition to a role in phloem transport of iron, nicotianamide (NA) is involved in iron homeostasis (Pich et al. 2001; Hell and Stephan, 2003). Immunochemical determination of NA in both cellular and extracellular components of leaves and root elongation zones in pea and tomato grown with varying levels of iron supply, as also in their iron-defective mutants, provide evidence to show chelation of excess iron accumulated in the vacuoles by NA (Pich et al. 2001).

Using castor bean seedlings model, Krüger et al. (2002) showed that phloem transport of iron occurs as a protein-iron complex. Essentially, all the [55]Fe fed to the seedlings was found to be associated with a protein fraction of the phloem. Purification of the iron transport protein (ITP) to homogeneity and cloning of the corresponding cDNA from the phloem exudates showed sequence similarity between ITP and the stress-related late embryogenesis-abundant protein family. It was also shown that the ITP binds preferentially to Fe^{3+} and not Fe^{2+} (Krüger et al. 2002).

Mobilization and translocation of iron in strategy I plants is influenced by iron accumulation in the apoplasmic fluid. Plants adapted to this strategy show substantial accumulation of iron in the apoplasm (apoplasmic pool). When plants are subjected to iron deficiency, the apoplastic iron is mobilized and translocated to the shoot (Bienfait et al. 1985; Longnecker and Welch, 1990). In a recent report, Graziano et al. 2002) suggested a role of nitric oxide (NO) in the internal distribution of iron and its delivery to active sites of metabolism. Supply of NO to iron-deficient plants prevented the development of chlorosis and inhibitors of NO accentuated chlorosis.

2.3 ROLE IN PLANTS

2.3.1 Enzyme Action

Iron enzymes and electron carrier proteins form integral components of mitochondrial and photosynthetic electron transport systems. A large number of iron enzymes, with a wide range of substrates, catalyze an array of reactions that are essential to primary metabolism and biosynthesis of secondary metabolites. They are involved in the biosynthesis of gibberellins, ethylene and jasmonic acid, which play important roles in regulation of developmental processes and also as signalling molecules. Iron may bind to proteins either in ionic form (non-haem iron enzymes) or in an organically bound form, as iron-protoporphyrin or haem prosthetic group (haem enzymes). In the non-haem enzymes, the iron ion forms a coordinate bond with the protein amino acids. A group of membrane-bound iron enzymes has recently been found to contain a coupled binuclear iron centre (Berthold and Stenmark, 2003). In these enzymes, two iron atoms forming an oxobridge, are bound to four carboxylate and two histidine residues (Fig. 2.2). An important example of the diiron enzymes is the mitochondrial alternative oxidase.

In a large group of non-haem iron enzymes, the iron cofactor is present in the form of an iron-sulphur (Fe-S) cluster, in which iron is coordinated to sulphide ions. Three main variants of iron-sulphur clusters are shown in Fig 2.3. Different aspects of Fe-S clusters have recently been reviewed by Johnson et al. (2005). Ease of inter-conversion between the two oxidation states of iron ($Fe^{2+} \leftrightarrow Fe^{3+}$) in the Fe-S cluster enables the Fe-S proteins to function as efficient acceptors or donors of electrons.

Iron-sulphur Clusters
Iron forms coordinate bonds with sulphide ions to forms three main type of iron-sulphur complexes (clusters):

(a) **Fe-S**: A single iron ion is tetrahedrally coordinated to the sulphydril groups of four cysteine residues of a protein.

(b) **2Fe-2S**: Two iron ions are coordinated to two inorganic sulphide ions.

(c) **4Fe-4S**: Contains four iron ions, four inorganic sulphide ions and four cysteine residues.

The Fe-S (single) clusters occur only in bacteria. Lack of one iron ion in a 4Fe-4S cluster forms 3Fe-4S clusters (succinate dehydrogenase). One of the iron ions of the iron-sulphur cluster undergoes a change in its oxidation state ($Fe^{3+} \leftrightarrows Fe^{2+}$). This enables the Fe-S cluster protein to function as acceptors or donors of electrons. Some non-haem iron enzymes, containing iron-sulphur clusters also contain an additional flavin (FMN or FAD) cofactor. Enzymes containing iron-sulphur and flavin prosthetic groups form key components of mitochondrial electron transport complexes.

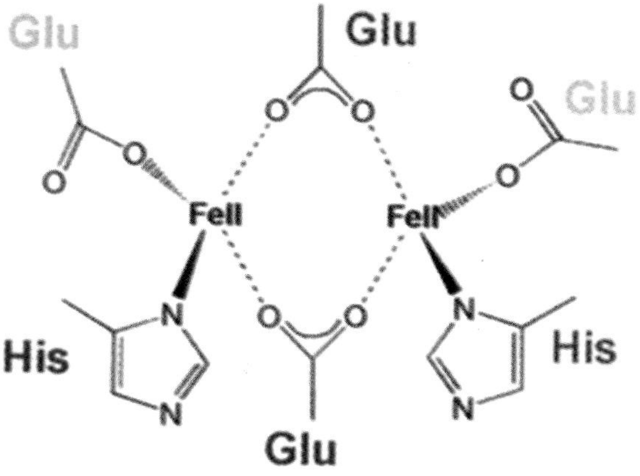

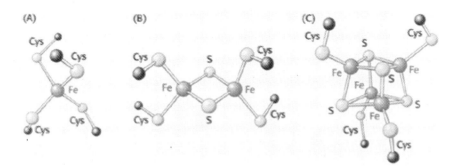

Fig. 2.2. Model of the coupled binuclear iron centre of diiron carboxylate proteins. The diiron center is liganded to four carboxylate (Glu) and two His residues. (Reprinted from *Annu.Rev.Plant Biol.* Vol. 54 ©2003, with permission from Annual Reviews. www. annualreviews.org.).

Fig. 2.3. Iron-sulphur clusters-(Fe-S) A, (2Fe-2S) B and (4Fe-S) C. In the (Fe-S) cluster Fe ion is tetrahedrally liganded to four Cys residues. In (2Fe-2S) clusters, the two Fe ions are bridged by sulphur ions. (Reprinted, with permission, from *Biochemistry.* JM Berg, JL Tymoczko, L Stryer, eds ©2002 W.H. Freeman and company.

In a group of iron enzymes, well known examples of which are catalase, peroxidases and cytochrome c oxidase, iron is bound to the apoprotein as iron porphyrin or haem, in which an iron ion is chelated to the four nitrogen atoms of the four pyrrole groups of protoporphyrin IX. Iron in the centre of the tetrapyrrole still possesses two additional ionic bonds (one on each side of the haem). Iron (usually Fe^{2+}) lies in the centre of the tetrapyrrole ring and undergoes oxidation-reduction (ferrohaem $\leftrightarrows$ ferrehaem).

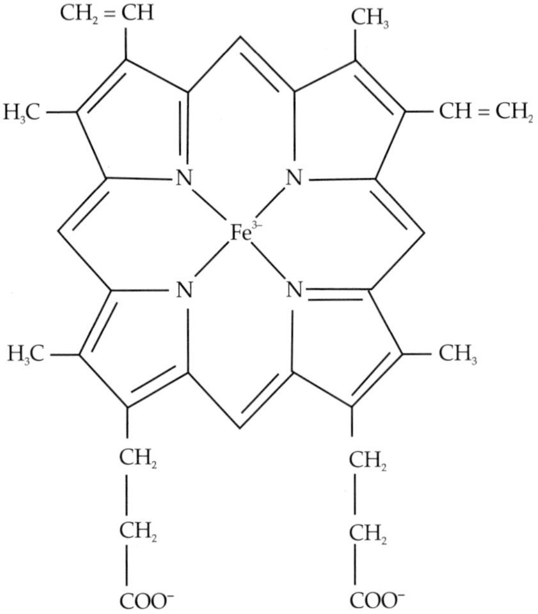

Haem b
(Iron-protoporphyrin IX)

Fig. 2.4. Haem (Fe-protoporphyrin IX). The Fe ion is coordinated to the four pyrrole nitrogen atoms. (Reprinted, with permission, from *Biochemistry*. D Voet, JG Voet, eds ©2004 John Wiley & Sons. Inc.).

2.3.1.1. Non-haem Iron Enzymes:

SUPEROXIDE DISMUTASE (EC 1.15.1.1)

The SODs belong to two separate families of enzymes with unrelated DNA sequences. One family contains an enzyme with Cu and Zn as metal constituents (Cu-Zn SOD). The other family contains two enzymes, one containing Mn (Mn-SOD) and the other containing Fe (Fe-SOD). The Fe-SOD contains a single Fe ion bound to an imidazole group along with a carboxyl group and is localized in the chloroplasts (Fridovich, 1986).

Superoxide dismutases (SODs) catalyze the dismutation of superoxide (O_2^-) ions to H_2O_2 and molecular oxygen (O_2).

$$2O_2^- + 2H^+ = O_2 + H_2O_2$$

The reaction is completed in two steps involving a cyclic change in the oxidative state of the metal (M). In the first step, the oxidized form of the metal (M_{ox}) reacts with one superoxide ion and a proton to form O_2 and gets reduced (M_{red}). In the second step, the reduced form of the enzyme reacts with another superoxide ion and a proton to form H_2O_2 and the oxidized form of the enzyme is regenerated. The enzyme uses two protons to convert two superoxide ions to generate a molecule of hydrogen peroxide.

$$M_{ox} + O_2^- + H \longrightarrow M_{red} + O_2$$
$$M_{red} + O_2^- + H^+ \longrightarrow M_{ox} + H_2O_2$$

LIPOXYGENASE (EC 1.13.11.12)

Lipoxygenases are non-haem containing fatty acid dioxygenases (Prescott and John,1996). They are monomeric, with each protein molecule containing a single iron (Fe^{2+}) atom. Lipoxygenases catalyze the introduction of O_2 into unsaturated fatty acids to form hydroperoxy derivatives. The soybean enzyme catalyzes the conversion of linoleic acid into 9 hydroperoxy-10,12 octadecanoic acid or 13 hydroperoxy-9,12- octadecanoic acid. This reaction can be summarized as:

$$\text{Linoleate} + O_2 \longrightarrow \begin{array}{l}\text{9-hydroxylinoleic acid/}\\ \text{13-hydroperoxy linoleic acid}\end{array}$$

The hydroperoxy derivatives of the polyunsaturated fatty acids are considered to play a key role in production of a variety of oxylipins, including jasmonate, which functions as a signalling compound in development and stress adaptation in plants (Feussner and Wasternack, 2002). Lipoxygenases are also involved in the production of free radicals (Siedow, 1991).

ALTERNATIVE OXIDASE

Alternate oxidase (AOX) activity, which was referred to as cyanide resistant respiration in earlier literature, is a non-haem iron enzyme, localized to the inner mitochondrial membranes (Siedow et al. 1995; Berthold et al. 2000). AOX has recently been shown to contain a coupled binuclear iron centre, with the iron (Fe^{2+}) atoms liganded to four carboxylate and two histidines residues (Berthold and Stenmark, 2003). When Cytochrome c oxidase is slowed down due to limited availability of ADP, the AOX functions as a non-proton pumping ubiquinol oxidase.

$$\text{Ubiquinol} + O_2 \longrightarrow \text{Ubiquinone} + H_2O$$

AOX functions as a shunt for the dissipation of excess reducing equivalents of the ubiquinol pool, preventing their interaction with molecular oxygen, which would lead to generation of reactive oxygen species (ROS). Thus,

AOX provides a mechanism for the overflow of excess reducing equivalents, minimizing the damage from ROS (Maxwell et al. 1999; Yip and Vanlerberghe, 2001).

2-OXOGLUTARATE-DEPENDENT ENZYMES

2-oxoglutarate-dependent dioxygenases (2-ODDs), which catalyze the hydroxylation, desaturation and epoxidation of a range of substrates using 2-oxyacids (e.g., 2-oxoglutarate) as a co-substrate, have a specific requirement of Fe^{2+} and, in many cases also, that of ascorbate. The 2-ODDs are involved in oxygenation and desaturation reactions in the biosynthetic pathways of flavonoids and signaling molecules (Prescott and John, 1996). They are also involved in hydroxylation of 2-deoxymugineic acid to 3-epihydroxy mugineic acid and 3-epihydroxy 2-deoxymugineic acid, which function as phytosiderophores.

2.3.1.2. Non-haem Iron Enzymes Containing Iron-Sulphur Cluster Cofactor

ACONITASE (EC 4.2.1.3)

Aconitase (Citrate [isocitrate] hydrolase) is a non-haem, iron-sulphur (4S-4S) protein. The enzyme contains four iron atoms complexed to four inorganic sulphides and three cysteine sulphur atoms, leaving one of the clusters free (for binding to the substrate [citrate, iso citrate]). The enzyme is localized in the cytosol and catalyzes the stereospecific dehydration-rehydration of citrate to isocitrate during the tricarboxylic acid (TCA) cycle. The reaction is catalyzed in two steps. The first step involves the dehydratation of citrate to *cis* aconitate and the second the hydratation of *cis* aconitate to isocitrate.

$$Citrate \rightleftharpoons Cis \text{ Aconitate} \rightleftharpoons Isocitrate$$

$$\overset{H_2O}{\nearrow} \qquad \overset{H_2O}{\searrow}$$

NITRITE REDUCTASE (EC 1.7.7.1)

Nitrite reductase (NiR) of higher plants is a monomeric protein containing two prosthetic groups: a [4Fe-4S] iron cluster and a sirohaem bridged by a sulphur ligand. The enzyme catalyzes six-electron reduction of nitrite (NO_2) to ammonium (NH_4^+)

$$NO_2^+ + 6 \text{ Fdx}_{red} + 8 \text{ H}^+ \longrightarrow NH_4^+ + 6 \text{ Fdx}_{ox} + 2H_2O$$

Nitrite reductase is localized in chloroplasts and root plastids. In both, the electrons for the reduction of nitrite are derived from ferredoxin (Fdx), but the chloroplastic and the root plastid ferredoxin differ in their source of reducing power. The chloroplastic Fdx is reduced by electrons ejected from P700 (non-cyclic electron transport), and the root plastid Fdx is reduced by NADPH produced during the oxidative pentose phosphate pathway, via ferredoxin:NADP reductase (FNR). This redox reaction is the exact opposite of the one taking place in chloroplasts and is made possible by the mid point potential of the plastid Fdx which is less negative than the chloroplast counterpart (Taniguchi et al. 1997).

SULPHITE REDUCTASE **(EC 1.8.7.1)**
Sulphite reductase is an iron-sulphur protein containing iron in the form of iron sulphur clusters. The enzyme catalyzes the six electron reduction of sulphite to sulphide, associated with conversion of 6 molecules of reduced ferredoxin (Fd_{red}) to their oxidative state (Fd_{ox}).

$$SO_3^{2-} + 6 \text{ Ferredoxin}_{red} \longrightarrow S_2 + 6 \text{ Ferredoxin}_{ox}$$

FORMATE DEHYDROGENASE **(EC 1.2.1.2)**
Formate dehydrogenase (FDH) catalyzes the oxidation of formate into carbon dioxide in the presence of NAD^+

$$\text{Formate} + NAD^+ = CO_2 + NADH$$

Higher plant FDH is essentially localized to the mitochondrial matrix in the non-green tissues. It has, however, been reported from *Arabidopsis* chloroplasts (Herman et al. 2002). Overexpression of FDH in response to abiotic stress conditions, particularly anaerobisis, enables the stressed plants to use formate as a respiratory substrate (Houston-Cabassa et al. 1998).

SUCCINATE DEHYDROGENASE **(EC 1.3.99.1)**
Succinate dehydrogenase (SDH) is an iron-sulphur flavoprotein. The prosthetic group of the enzyme contains three types of iron-sulphur clusters: 2Fe-2S, 3Fe-4S and 4Fe-4S. The flavin component is the oxidized form of flavin adenine dinucleotide (FAD). The enzyme catalyzes the oxidation of succinate to formate with concomitant reduction of FAD to $FADH_2$, which still remains bound to the enzyme.

$$\text{Succinate} + FAD \longrightarrow \text{Fumarate} + FADH_2$$

The enzyme links the citric acid cycle to the mitochondrial electron transport chain, by passing on the two electrons from $FADH_2$ to the quinol pool of the ETC through Succinate - Q reductase and through it to molecular oxygen (O_2) (Section 1.3.3.).

NADH - Q OXIDOREDUCTASE **(EC 1.6.99.5)**
NADH - Q oxidoreductase (NADH dehydrogenase) is a high molecular weight ($\geq$ 88 kd) protein containing at least 34 polypeptide chains. Like SDH, NADH-Q oxidoreductase is also an iron flavoprotein, but in NADH-Q oxidoreductase, the flavin component is an oxidized flavin

mononucleotide (FMN) and a series of Fe-S type iron clusters. NADH-Q oxidoreductase catalyzes the reaction:

$$NADH + Q + 5H^+_{matrix} \longrightarrow NAD^+ + QH_2 + 4H^+_{Cytosol}$$

The FMN prosthetic group of the enzyme accepts high-potential electrons from NADH and moves them through the series of iron sulphur clusters ending up in the reduction of FMN to $FMNH_2$. As a part of the enzyme molecule is embedded in the mitochondrial matrix and a part protrudes into the cytosol, it also acts as a proton pump.

FDX – THIOREDOXIN REDUCTASE

Ferredoxin-thioredoxin reductase (FTR) contains iron in the form of [4Fe-4S] cluster closely associated to redox active disulphide of the protein. Localized to chloroplasts, FTR catalyzes two one-electron oxidations of ferredoxin and in doing so, its disulphide bond is reduced to sulphydryl groups. The reduced form of the enzyme plays important regulatory roles in chloroplasts (Kim and Mayfield, 1997; Schürmann and Jacquot, 2000). Reduced FTR donates electrons to chloroplast protein disulphide isomerase (cPDI) which, in turn, reduces chloroplast polyadenylate-binding protein (cPABP). Reduced cPABP binds to *psb*A mRNA causing enhanced translation of the D1 protein in light (Kim and Mayfield, 1997). FTR catalyzed reduction of the stromal thioredoxins forms an efficient mechanism (ferredoxin-thioredoxin system) for the regulation of disulphide containing chloroplast enzymes involved in Calvin cycle and degradative processes. The Calvin cycle enzymes, which in the oxidized state are inactive (in the dark), are reduced to an active form (in light); and the degradative enzymes, such as fructose-6-phosphate dehydrogenase, which are active in the oxidized form (in the dark) are reduced to their inactive form. In doing so, FTR contributes to light regulated separation of biosynthetic and catabolic processes, minimizing their simultaneous operation.

2.3.1.3 Haem Enzymes

CATALASE (EC 1.11.1.6)

Catalase (CAT) is a haem enzyme. It contains four iron atoms per enzyme molecule, chelated to protoporphyrin IX. The enzyme catalyzes the breakdown of hydrogen peroxide (H_2O_2) into water (H_2O) and molecular oxygen at a very fast rate.

$$2H_2O_2 \rightarrow 2H_2O + O_2$$

As an enzymic constituent of the antioxidative defense system, catalase acts as a scavenger of H_2O_2 generated by the SOD catalyzed dismutation of superoxide ions (O_2^-). The enzyme is highly concentrated in peroxisomes, where it plays a protective role. It protects the cellular constituents (including itself) from toxic effects of H_2O_2 generated during the conversion of glycolate to glyoxylate

$$\text{Glycolate} \xrightarrow[\quad O_2 \qquad H_2O_2 \quad]{} \text{Glyoxylate}$$

Catalase also shows peroxidative activity by using H_2O_2 to oxidize substrates such as ethanol to acetaldehyde:

$$H_2O_2 + CH_3CH_2OH \rightarrow 2H_2O + CH_3CHO$$

PEROXIDASES (EC 1.11.1.7)

Peroxidases (POXs) are haem-containing glycoproteins encoded by a large multigene family. Peroxidases generally lack donor specificity. They catalyze the dismutation of hydrogen peroxide to water using a wide range of donors.

$$AH_2 + H_2O \rightarrow A + H_2O$$

Ascorbic peroxidase utilizes ascorbic acid as its specific electron donor to reduce H_2O_2 to water with concomitant generation of monodehydro-ascorbate. Peroxidases also catalyze the oxidation of alcoholic (phenolic) compounds producing free radical intermediates, which polymerize and cross-link the cell wall constituents of the plants. Peroxidases bound to cell walls play a role in cell wall lignification.

Several white rot fungi peroxidases function as manganese peroxidases, catalyzing the oxidation of Mn (II) to Mn (III) by hydrogen peroxide (Gold and Alic, 1993). The manganese peroxidases possibly posses two binding sites that differ in affinity for Mn (II) (Mauk et al. 1998). Mn (III), resulting from the oxidation of Mn (II), catalyzes the oxidative degradation of lignin.

CYTOCHROME c OXIDASE (EC 1.9.3.1)

Cytochrome c oxidase is a haem-copper protein. The enzyme contains two haem molecules (haem a and haem a_3), and two copper centres, one containing two copper ions (Cu_A / Cu_A) and the other a single copper ion (Cu B), bound to a 13 polypeptide chain. The enzyme is universal and localized in the mitochondrial membrane, where it catalyzes the regeneration of reduced cytochrome c, concomitant with the reduction of molecular oxygen to water. The Cu_A/Cu_A site acts as the acceptor of electrons from reduced cytochrome c. The proximity of haem a_3 and Cu_B enables them to form an active centre for the reduction of O_2 and the transfer of eight protons from the mitochondrial matrix. Four of these are used up in the regeneration of four molecules of reduced cytochrome c oxidase (in the membrane) and four in the reduction of O_2 to $2H_2O$. These reactions are achieved through cyclic reactions involving four molecules of cytochrome c oxidase and can be summarized as:

$$4\text{Cyt } c_{red} + 8\text{H}^+_{matrix} + \text{O}_2 \longrightarrow 4\text{CytI } c_{ox} + 2\text{H}_2\text{O} + 4\text{H}^+_{cytosol}$$

SUCCINATE-Q REDUCTASE COMPLEX

Succinate-Q reductase (mitochondrial electron transport chain complex II) is an iron-flavoprotein enzyme. The iron component is a 4Fe-4S type cluster and the flavin component is a molecule of FAD. The enzyme is ubiquitous and localized to the mitochondrial membrane. The Fe-S centre of the enzyme accepts the electrons generated during the oxidation of succinate to fumarate (citric acid cycle) and transports them to ubiquinone (Q) for entry into the mitochondrial electron transport chain.

$$\text{FADH}_2 \xrightarrow{e^-} \text{Fe-S} \xrightarrow{e^-} \text{Q} \xrightarrow{e^-} \text{QH}_2$$

Q - CYTOCHROME c OXIDOREDUCTASE (EC 1.9.99.1)

Q - cytochrome c oxidoreductase (Cytochrome bc_1 complex, complex III) is a haem-protein. It is a dimer, with each monomer containing multiple subunits. The enzyme contains three molecules of haem and a 2Fe-2S type ion-sulphur cluster. Also known as **Reiske** centre, the enzyme contains two binding sites for ubiquinone (Q_o and Q_i). The enzyme catalyzes the transfer of electrons from ubiquinol (QH_2) to oxidized cytochrome c and pumps out protons from the mitochondrial side of the matrix into the cytosol, regenerating ubiquinone (Q).

$$\text{QH}_2 + 2\text{cyt } c_{ox} + 2\text{H}^+ \longrightarrow \text{Q} + 2\text{cyt } c_{red} + 4\text{H}^+$$

Cytochrome P450 Monooxygenases

A group of monooxygenases that catalyze the incorporation of one atom of oxygen from O_2 into the substrate and reduce the other to water belong to a superfamily of haem-containing proteins which require cytochrome P450 for catalysis. They are known as cytochrome P450 monooxygenases, simply P450 monooxygenases or haem-thiolate proteins. The cytochrome P450 monooxygenases catalyze the oxidation or hydroxylation of a wide range of substrates using NADPH as a co-substrate. With a few exceptions, cytochrome P450s do not directly react with NADPH. Electrons from NADPH are first transferred to a Fe-S flavoprotein—known as P450 reductase—then to cytochrome P450 and finally to the haem protein. Beginning with the cloning of ripening-related cytochrome P450 gene from avocado fruits (Bozak et al. 1990), many membrane (ER) localized cytochrome P450 encoding genes (designated CYP for cytochrome P450) have been cloned and characterized. Enzymes coded by the CYP genes have since been shown to be involved in fatty acid metabolism, biosynthesis of secondary metabolites, plant hormones and signalling molecules (Chapple, 1998). Some cytochrome P450 monooxygenases of higher plants have been listed in Table 2.1. Functional

Table 2.1. Cytochrome P450 dependent monooxygenases of higher plants

Enzyme	Reaction catalyzed	Role
Allene oxide synthase (AOS) (hydroperoxide dehydratase)	13-Hydroperoxylinolenic acid ⟶ Allene oxide	Fatty acid metabolism; JA biosynthesis
Fatty acid Hydroperoxide lyase (HPO Lyase)	13-Hydroperoxylinolenic acid ⟶ hex-3-enol + dodec-9-enoic acid-12al	Fatty acid metabolism; Fruit and vegetable volatiles
Tyrosine-N-hydroxylase($P450_{TYR}$) (a multifunctional enzyme)	Tyrosine ⟶ p-hydroxy phenyl acetaldoxime p-hydroxyphenyl acetaldoxime ⟶ p-hydroxy mandelonitrile	Fatty acid metabolism; Biosynthesis of cyanogenic glucosides
ent-kaurene oxidase (KO)	ent kaurene ⟶ ent-kaurenol ⟶ ent-kaurenal ⟶ ent-kaurenoic acid	Gibberellin biosynthesis
Cinnamate-4-hydroxylase (C4H)	$Trans$-cinnamic acid ⟶ p-coumaric acid	Phenyl propanoid pathway; Biosynthesis of secy. metabolites (lignin, flavonoids, stelbenes)
Ferrulate-5-hydroxylase (F5H)	Ferrulic acid ⟶ 5-hydroxy ferrulic acid	Lignin biosynthesis (*Arabidopsis*)
Flavonoid – 3′5′-hydroxylase (F3′5H)	Naringenin ⟶ Eryodictyol + Pentahydroxyflavonone also: Dihydrokempferol ⟶ Dihydroquercitin + Dihydromyricitin	Anthocyanin biosynthesis

analyses of cloned genes and their expression in transgenics (Schuler and Werck-Reichart, 2003) have shown that some reactions catalyzed by the cytochrome P450 monooxygenases may be rate-limiting steps in the biosynthesis of plant hormones.

2.3.2 Iron Proteins

Several iron proteins function as carriers of electrons, forming part of the mitochondrial and chloroplastic electron transport systems (Section 1.3.3.). These include the haem proteins (*cytochromes*) as well as iron-sulphur (2Fe-2S) proteins (*ferredoxins*). The root nodules of leguminous plants contain a haemoglobin type protein *leghaemoglobin*, with a very high affinity for molecular oxygen. Another protein, *phytoferritin* serves as a major store for cells' excess iron in plants.

CYTOCHROMES

Cytochromes are haem proteins, which function in the transfer of electrons by undergoing reversible change in the oxidative state of iron ($Fe^{2+} \leftrightarrows Fe^{3+}$). They are localized in mitochondrial membranes, chloroplasts and in endoplasmic reticulum. The mitochondrial electron transport system contains cytochromes b, c_1, c and a_3. Cytochrome b contains two subunits, Cyt b_L and Cyt b_H, which differ in affinity of the haem (L-low, H-high). Cytochromes c_1 and c differ from other cytochromes in the sense that their iron porphyrin group is covalently linked to the protein. Cytochromes c and a_3 together constitute cytochrome c oxidase which catalyzes the reduction of molecular oxygen to water in the terminal reaction of the mitochondrial electron transport system.

The chloroplasts contain a Cyt b_6f complex, which catalyzes the transport of electrons from plastoquinone to the copper protein plastocyanin. The Cyt b_6f complex contains at least three components: one containing two b-type haems (Cyt b_6), another is the Reiske type Fe-S protein; and another one is a molecule of cytochrome f, which is a c-type cytochrome. The Cyt b_6f complex is also involved in the transfer of electrons during the cyclic electron flow in photosystem I (Fig. 1.3). Instead of moving electrons to Ferredoxin-NADP reductase, reduced ferredoxin moves the electrons to cytochrome b_6f complex. Since electron movement through the cytochrome b_6f complex involves spontaneous proton pumping into the thylakoid lumen, cyclic electron transport adds to a proton gradient. This assumes great importance when the requirement of ATP is more than that for NADPH. Reduced cytochrome b_6f complex moves the electrons back to reduced plastocyanin, which is regenerated on absorption of light by P700. A b type cytochrome, cytochrome b_6, is also involved in catalyzing the desaturation of fatty acids in endoplasmic reticulum.

LEGHAEMOGLOBIN

Leghaemoglobin is an oxygen-binding protein. The iron component of the protein is structurally similar to the haem of the haem enzymes and cytochromes. Leghaemoglobin has high affinity for O_2 and regulates its concentration in the root nodules of leguminous plants to a critical level, which is sufficient to meet the growth requirement of the bacteroids, but not inhibitory to nitrogenase. Nitrogenase is strongly inhibited by O_2. Binding of O_2 to leghaemoglobin, in preference to nitrogenase, prevents the enzyme from inactivation, allowing it to fix molecular nitrogen. Roots of non-nodulating plants are also reported to contain small amounts of leghaemoglobin (Appelby et al. 1988), which possibly triggers a shift from aerobic respiration to fermentation, on sensing deficiency of O_2 (signal transduction).

FERREDOXIN

Ferredoxins (Fdx) are low molecular weight (9 kDa), water-soluble iron proteins. They possess [2Fe-2S] type iron-sulphur clusters and function in transport of electrons by a reversible change in the oxidative state of the iron ion ($Fe^{3+} \rightleftharpoons Fe^{2+}$). The ferredoxins are negatively charged and, on transfer of electrons to other proteins, form electrostatistically stable complexes. They are mainly localized in leaf chloroplasts and root plastids. The chloroplastic and plastid ferredoxins differ in their electronegativity. The mid-point potential of the root plastid Fdx is substantially lower (-320 mV) than the chloroplastic Fdx (-387 mV) (Taniguchi et al. 1997).

The chloroplastic feredoxin, reduced on accepting the electrons ejected by P700, functions as an electron donor for several reductive reactions. In a reaction mediated by ferredoxin-NADP$^+$ reductase (FNR), it catalyzes the reduction of NADP$^+$ to NADPH, which functions as a reductant for Calvin cycle and many other biosynthetic reactions. Alternatively, it may transfer the electrons to ferredoxin-thioredoxin reductase (FTR) on the way to thioredoxin, reduced form of which regulates the activities of several enzymes. Reduction of the disulphide bond to sulphydryl form by reduced thioredoxin activates several enzymes of the Calvin cycle (e.g. Fructose 1,6-biphosphatase, sedoheptulose 1,7-biphosphatase, phosphoribulose kinase, NADP$^+$-glyceraldehyde 3-phosphate dehydrogenase) and inactivates some degradative enzymes such as glucose-6-phosphate dehydrogenase (Schürmann and Jacquot, 2000). Reduced ferredoxin also serves as the immediate source of electrons for the reduction of nitrite and sulphite and reductive deamination of glutamine to glutamate.

FERRITIN

Ferritins are non-haem proteins involved in iron storage (Theil, 1987). Holoprotein is a shell, which defines a central cavity that may accommodate upto 4500 iron atoms (17 to 25% iron). The plant ferritins, *phytoferritins*, form a major sink for storage of iron (Seckbach, 1982; Smith, 1984) and, subsequently, its release for cellular functions. Synthesis of ferritin is regulated by iron status of plants (Lobreaux et al. 1992). Induction of ferritin synthesis in response to iron overload (excess) forms a mechanism for reducing the concentration of free iron, an excess of which may accelerate free radical reactions (Halliwell and Gutterridge, 1986; Price and Hendry, 1991), causing lipid peroxidation, protein degradation and mutagenesis. Effective protection against oxidative damage, triggered in response to iron overload, is provided by inductive synthesis of ferritin through overexpression of ferritin mRNA (Van der Mark et al. 1983; Lescure et al. 1991; Lobreaux et al. 1992, 1995). Exposure of cultured soybean cells to excess of iron causes activation of the ferritin gene (Lescure et al. 1991), substantiating the fact that ferritin synthesis in plants is controlled at the transcriptional level. Maize mutant YS1, which accumulates less iron than the wild type plants, shows enhanced accumulation of ferritin m RNA on treatment with iron (Fobis-Loisy et al. 1996).

2.3.3 Mitochondrial Electron Transport

Iron is a constituent of a series of electron transport proteins of mitochondrial membranes. These iron proteins function in the transfer of electrons from NADH and $FADH_2$ (produced during the citric acid cycle) to molecular oxygen. The mitochondrial electron transport chain and the sequence of carriers therein are shown in Figure 1.2. The transfer of electrons from NADH through the components of the electron transport chain is accompanied by generation of a proton motive force, which is transduced to synthesize ATP (oxidative phosphorylation).

2.3.4 Photosynthesis

Photosynthetic Electron Transport
Iron plays an important function in photosynthetic electron transport (Fig 1.3). The reaction centre of photosystem II contains iron in the ionic form and that of photosystem I as three molecules of 4Fe-4S type iron-sulphur clusters. During electron transport through photosystem II, electrons move from P680 to pheophytin and then to plastoquinone (Q), reducing it to plastoquinol (QH_2). The reoxidation of QH_2 and further movement of electrons, ultimately ending in reduction of NADP, involves movement of electrons through the components of photosystem I. The link between quinol in photosystem II and the photoreceptor complex of photosystem I (P700) is provided by the cytochrome b_6f complex and the copper protein plastocyanin (PC).

$$QH_2 + 2PC\ (Cu^{++}) \longrightarrow Q + 2PC\ (Cu^+) + 2H^+$$

On absorption of light by the photoreceptor pigments of the photosystem I (P700), the electrons derived from reduced Plastocyanin (PC_{red}) move to P700 (Chl a) and, subsequently, through A_0 (a phylloquinone binding protein), A_1 (Vit K1) and three [4Fe-4S] iron sulphur clusters to ferredoxin. The transfer of electrons from reduced ferredoxin (Fdx_{red}) to $NADP^+$ is catalyzed by Fdx-NADP$^+$ reductase (FNR). In a two-step reaction, the FAD prosthetic group of the Fdx-NADP$^+$ reductase accepts two electrons from two molecules of reduced ferredoxin—one at a time—and transfers a hydride ion and a proton to $NADP^+$ to produce NADPH, which serves as a reductant in Calvin cycle and in many other biosynthetic reactions.

$$2Fdx_{red} + NADP^+ + H^+ \longrightarrow 2Fdx_{ox} + NADPH$$

CALVIN CYCLE
Iron, in the form of iron-sulphur clusters, is an integral part of ferredoxin and ferredoxin-thioredoxin reductase involved in the photoreduction of

thioredoxin (Droux, 1989). The ferredoxin-thioredoxin system provides a redox mechanism for the regulation of disulphide containing enzymes of the Calvin cycle (Schürmann and Jacquot, 2000).

The chloroplastic thioredoxins reduced in response to the light signal from PS1 activates fructose 1,6-biphosphatase, sedoheptulose 1,7-biphosphatase, ribulose 5-phosphokinase and $NADP^+$-glyceraldehyde 3-phosphate dehydrogenase by reducing their disulphide bonds to sulphy-dryl groups. In the dark, when the photoreduced thioredoxin is converted back to its oxidized state, these enzymes are spontaneously converted to their inactive or sub-active form. In some plants, the ferredoxin-thioredoxin system also regulates the activity of rubisco activase, thus, activating Rubisco (Zhang and Portis, 1999). Hence the iron-sulphur clusters function as redox signals for light-dark regulation of Calvin cycle (Dai et al. 2000).

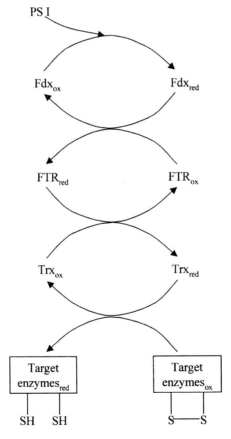

Fig. 2.5. The ferredoxin—thioredoxin system

2.3.5 Fatty Acid Metabolism

The reduction of ferredoxin by electrons drawn from water molecule during photosystem II serves as a source of high reducing power required for the synthesis of fatty acids. Repeated carboxylation and acylation of acetyl CoA and malonyl CoA produce saturated (C_{16} and C_{18}) fatty acids. Three membrane bound iron proteins—**NADH-cytochrome b_5 reductase, cytochrome b_5**—and a **desaturase** are involved in desaturation of the fatty acids (Fig. 2.6), providing fluidity to the membrane lipids. The desaturase uses O_2 and NADH or NADPH to introduce a double bond in the fatty acids.

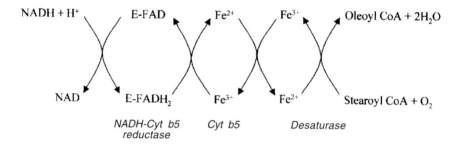

Fig. 2.6. Electron transport chain in the desaturation of fatty acids (After Berg et al. 2001)

At first, electrons are transferred from NADH to FAD moiety of NADH-cytochrome b_5 reductase. The haem iron atom of cytochrome b_5 is then reduced to Fe^{2+} state. The non-haem iron atom of the desaturase is subsequently converted into the Fe^{2+} state, which enables it to interact with O_2 and the saturated fatty acyl CoA structure. A double bond is formed and two molecules of H_2O are released.

2.3.6 Growth Hormones and Signalling Molecules

Iron is involved in biosynthesis of plant hormones and signalling molecules in a big way. Enzymes with iron cofactors catalyze reactions in the biosynthetic pathways of gibberellins, ethylene and jasmonic acid. Conversion of *ent*-kaurene to *ent*-kaurenoic acid involves three sequential cyclization reactions catalyzed by *ent*-kaurene oxidase, which is a cytochrome P450 monooxygenase (Helliwell et al. 2001). Molecular cloning and functional expression of gibberellin oxidases from *Arabidopsis* has revealed that several steps in the biosynthesis of gibberellins from GA_{12} aldehyde are catalyzed by 2-oxoacid dependent non-haem dioxygenases (Hedden and Phillips, 2000). These include the GA20 oxidase and GA3 oxidase. The GA oxidases have also been shown to be involved in the conversion of active GAs to their inactive forms. Overexpression of *Arabidopsis* GA_{12} oxidase has been shown to result in dwarfing of plants as a result of lowering of GA concentration (Schomburg et al. 2002).

The terminal step of ethylene biosynthesis involving oxidation of 1-aminocyclopropane –1–carboxylic acid (ACC) to ethylene is catalyzed by the non-haem iron enzyme ACC oxidase. The enzyme contains Fe^{2+} as the metal cofactor (Bouzayen et al. 1990, 1991). Biosynthesis of jasmonic acid from linoleic acid involves activities of the non-haem iron enzyme lipoxygenase and two cytochrome P450 dependent monooxygenases—fatty acid hydroperoxide lyase and allene oxide synthase (AOS) (Howe et al. 2000; Feussner and Wasternack, 2002). Conversion of 13-hydroperoxylinoleic acid and that of allene oxide by AOS, are rate-limiting steps of jasmonic acid biosynthesis. Overexpression of AOS gene in transgenics leads to enhanced synthesis of jasmonic acid (Harms et al. 1995). Cytochrome P450s have also been involved in the tryptophan-dependent, indole-3-acetonitrile pathway of IAA biosynthesis. Hull et al. (2000) have cloned and characterized two *Arabidopsis* genes *CYP 79 B2* and *CYP 79 B3* which encode the two P450 enzymes that catalyze the conversion of tryptophan to indole-3-acetaldoxine which, in turn, is converted to indole-3-acetonitrile, the immediate precursor of IAA. Indole-3-acetonitrile is also considered a likely precursor of glucosinolates. The cytochrome P450s are also involved in the biosynthesis of brassinosteroids (Crozier et al. 2000).

The non-haem dioxygenases are involved in the biosynthesis of flavanoids and anthocyanin (Prescott and John, 1996). Flavanones are converted to flavonals involving two enzyme-catalyzed steps. In the first reaction, flavanone is converted to dihydroxyflavonal by flavone-3'-5'-hydroxylase. In the second reaction, dihydroxyflavonal is converted to flavonal by flavonal synthase (FS). Both the enzymes have a specific requirement of Fe^{2+} for activation. Dihydroxyflavonal also serves as the branching point for the synthesis of anthocyanins.

2.3.7 Free Radical Production

The superoxide ions (O_2^-) generated out of one-step reduction of oxygen (O_2) and hydrogen peroxide (H_2O_2), to which O_2^- are dismutated by the superoxide dismutases, make two reactive oxygen species that cause oxidative damage to cellular constituents. Allowed to react together, O_2^- and H_2O_2 produce still more toxic hydroxyl radicals (Haber-Weiss reaction). This reaction is catalyzed by free and loosely bound ionic iron. Free Fe^{2+} reacts with H_2O_2 to produce hydroxyl radicals (Fenton reaction)

$$Fe^{2+} + H_2O_2 \longrightarrow Fe^{3+} + OH^\bullet + OH^-$$

The Fe^{3+} produced by Fenton reaction, on reacting with superoxide ions, is cycled back to Fe^{2+}

$$Fe^{3+} + O_2^- \longrightarrow O_2 + Fe^{2+}$$

Thus, traces of free inorganic iron accelerate the Haber-Weiss reaction, generating highly toxic hydroxyl radicals and resulting in oxidative damage to diverse cellular constituents (Halliwell and Gutteridge, 1989).

$$O_2^- + H_2O_2 \longrightarrow OH^\bullet + OH^- + O_2 \text{ (Haber-Weiss reaction)}$$

Fe^{2+} may also react with molecular oxygen (O_2) to produce still more toxic compounds such as ferryl ($Fe^{2+}O$) or perferryl ($Fe^{2+}O_2$) (Qian and Buettnov, 1999). These highly reactive oxygen species damage membrane lipids, proteins and DNA and induce mutations. Another way iron could contribute to production of reactive oxygen species is through action of lipoxygenase. This iron enzyme, essentially involved in hydroxylation of linoleic acid, may also catalyze the production of singlet oxygen species ($^1O_2$).

2.3.8 Protective Role

Iron plays a role in protection of plants against damage from reactive oxygen species, salt stress and pathogen attack.

DETOXIFICATION OF REACTIVE OXYGEN SPECIES

While free ionic iron accelerates the generation of reactive oxygen species (Halliwell and Gutteridge, 1989; Becana et al. 1998), iron in the protein environment (enzymes) contributes to their effective detoxification. Iron deficiency may affect both the generation of reactive species and the antioxidative defenses (Iturbe-ormaetxe et al. 1995; Becana et al. 1998). The iron isoenzymes of superoxide dismutase (Fe-SOD) carry out the detoxification of O_2^- by dismutating them to H_2O_2. The haem enzymes catalase and ascorbate peroxidase acts as effective scavengers of H_2O_2 in the different cellular compartments. Iron deprivation of sunflower plants is reported to result in marked decrease in the activity of ascorbate peroxidase, concomitant with increased accumulation of hydrogen peroxide (Ranieri et al. 2001). As a component of the mitochondrial electron transport chain, the iron enzyme alternative oxidase (AOX) provides an alternate pathway to the reducing equivalents from quinol and prevents them from interacting with oxygen to generate reactive oxygen species (Purvis, 1997; Millenaar et al. 1996, 1999).

OSMOPROTECTION

As a Reiske type cofactor of Fdx-Choline monooxygenase (Rathinasabapathy et al. 1997), iron is involved in the biosynthesis of glycinebetaine, which offers osmoprotection against high salinity and extremes of temperature by maintaining the structure and integrity of plasma membrane (Gorham, 1995, references therein).

Some iron enzymes are involved in defense against plant pathogens. Haem peroxidases are suggested to be involved in polymerization of the cinnamyl alcohols to form lignin (Hartfield and Vermerris, 2001), which contributes to mechanical resistance to pathogen attack. Caruso et al. (2001) have shown a more direct involvement of haem peroxidase in defense against invading pathogens. Caruso et al. (2001) purified and characterized a peroxidase from wheat kernel and showed it to inhibit the germ tube elongation of phytopathogenic fungi.

2.4 DEFICIENCY RESPONSES

2.4.1 Visible Symptoms

Plants subjected to iron deficiency develop visible symptoms considerably early, even before growth becomes depressed. Deficiency typically appears in the form of chlorosis of young leaves, appearing first at the base of leaves/leaflets. Chlorosis gradually intensifies in the interveinal areas, appearing as chlorotic striping in parallel veined leaves (in graminaceous plants) and fine chlorotic mottling or 'marbling' in reticulate veined leaves. Prolonged or acute deficiency leads to bleaching and disintegration of mesophyll of the chlorotic areas, as a result of which these areas become dry and papery. The severely chlorotic areas may turn brown and necrotic. Some plants may develop purple or pink tints.

Plants vary in sensitivity to iron deficiency. Amongst cereals, compared to wheat, barley and oats, maize, rice and sorghum are more susceptible to iron deficiency. Even at the seedling (nursery) stage, rice develops severe chlorosis. Apical part of young leaves of sorghum plants turns severely chlorotic and exudes a viscous fluid, indicating leakage of solutes through the membranes of epidermal cells. The chlorotic leaves may also develop reddish brown necrotic lesions. When subjected to severe deficiency of iron, the coloured petals of some crucifers and legumes show loss of pigmentation, which agrees with the involvement of cytochrome P450 monooxygenases in the biosynthesis of flavanoids (section 2.3).

In fruit trees, iron deficiency is evident in the reduction in leaf size and chlorosis of young leaves. The severely chlorotic lamina turns papery and leaf margins often turn reddish brown. Under severe deficiency, young shoots may even 'die-back'.

2.4.2 Enzyme Activities

Activities of several enzymes have been studied in relation to iron nutrition of plants. Amongst the iron enzymes investigated in response to iron deficiency, the haem enzymes-catalase and peroxidase figure prominently. There are several reports of iron-deficient plants showing decreased activities of catalase and peroxidase (Agarwala and Sharma, 1961; Agarwala et al. 1965; Machold, 1968; Del Rio et al. 1978a; Bisht et al. 2002a). Sijmons et al. (1985) have described low activity of suberin specific peroxidases in roots of iron-deficient beans. Catalase has been found to be more responsive to iron-deficiency than peroxidase (Agarwala et al. 1965; Machold, 1968; Bisht et al. 2002a). Low activity of catalase is also obtained in plants showing lime-induced chlorosis (Bertamini et al. 2002) or when they are subjected to excess supply of heavy metals (Agarwala et al. 1977a; Pandey and Sharma, 2002b, 2003). While a decrease in catalase activity is a universal response to iron deficiency, it is not always so in case of peroxidase. Instances of iron deficiency induced increase in peroxidase activity are also known (Dekock et al. 1960). Species specificity and developmental stage of plants may contribute to differences in peroxidase-related responses of iron-deficient plants. Ranieri et al. (2001) made a study of the effect of iron deficiency on the activities of non-specific peroxidase, as well as ascorbate-specific peroxidase (involved in the detoxification of hydrogen peroxide) and found that iron deprivation of sunflower plants made little difference to the activities of the non-specific peroxidase but produced a marked decrease in the activity of ascorbate peroxidase. Other iron enzymes—activities of which are reported to be low in iron-deficient plants—include aconitate hydratase (Hsu and Miller, 1968; Mehrotra and Jain, 1996) and Fe-SOD (Leidi et al. 1987a). Formate dehydrogenase, a key enzyme of anaerobic metabolism, is reported to be overexpressed in response to iron deficiency (Suzuki et al. 1998).

Iron nutrient status also influences the activities of enzymes of which it is not a constituent or a specific activator. The iron-deficiency response of these enzymes includes activation as well as inhibition of enzyme activities. Rabotti and Zocchi (1994) showed a decrease in the activity of H^+ ATPase, involved in iron reduction, in the roots of cucumber plants exposed to iron deficiency. Espen et al. (2000) reported enhancement in activities of glyceraldehyde-3 phosphate dehydrogenase, pyruvate kinase and fructose 6-phosphate kinase in roots of Fe-deficient cucumber plants, which suggests accelerated glycolysis with concomitant decrease in starch concentration. Plank et al. (2001) described iron-dependent changes in the activity of acetyl CoA carboxylase. The activity of acetyl CoA carboxylase in leaves of soybean plants showed a decrease in response to iron deficiency. Several enzymes have been shown to be activated when plants are grown under conditions of iron deficiency. Tomato leaf acid phosphatase shows several fold increase on iron deprivation (Hewitt and Tatham, 1960; Bisht et al. 2002a). Bisht et al. (2002a) showed a reversal of enhancement in enzyme activity following increase in iron supply from deficiency to sufficiency. Like acid phosphatase, ribonuclease is also activated under iron deficiency (Agarwala et al. 1965; Bisht et al. 2002a). Iron-induced activation of ribonuclease is also reversed on recovery from iron deficiency (Bisht et al. 2002a). Phospho*enol* pyruvate carboxylase (PEPCase) also shows a marked increase in response to iron deficiency (Rabbotti et al. 1995; De Nisi and Zocchi, 2000). De Nisi and Zocchi (2000) have suggested a role of iron in regulation of PEPCase activity at the transcriptional level.

2.4.3 Photosynthesis

Involvement of iron in chloroplast development, harvesting of light energy and transport of electrons from water to $NADP^+$ leads to impairment of photosynthesis under iron deficiency. Chloroplasts of iron-deficient plants are smaller in size and show a decrease in the number of thylakoid membranes and an increase in stroma (Spiller and Terry, 1980; Platt-Aloia et al. 1983). Thylakoids show a marked decrease in lipid and protein content (Nishio et al. 1985a). Iron deficiency induced decrease in chloroplastic proteins is much more than in cytoplasmic proteins (Shetty and Miller, 1966). Chloroplasts of iron-deficient plants also lack in polyribosomes (Lin and Stocking, 1978).

Chloroplasts of iron-deficient plants show a decrease in the concentration of light harvesting pigments. There are several reports of decrease in chlorophyll concentration in plants suffering from iron deficiency (Marsh et al. 1963; Hsu and Miller, 1968; Machold, 1968; Del Rio et al. 1978a;Miller et al. 1984; Spiller et al. 1982; Chereskin and Castelfranco, 1982; Leidi et al. 1987b). Barley plants grown with varying levels of iron showed close correspondence between the levels of iron supply and leaf chlorophyll concentration (Agarwala and Sharma, 1961). Decrease in chlorophyll concentration on iron deprivation is reversed on resumption of adequate supply of iron (Agarwala et al. 1965; Bisht et al. 2002a). Terry and Low (1982) demonstrated the effect of iron deficiency on chlorophyll to be associated with changes in the intracellular localization of iron. Iron-deficient plants also exhibit changes in the concentrations of carotenoids and xanthophylls (Morales et al. 1990, 1992; Quilez et al. 1992; Abadia, 1992). Concentrations of the individual carotenoids and xanthophylls may, however, be affected differentially. Characteristically, iron-deficient leaves show decrease in the concentration of β-carotene and neoxanthine and increase in the concentration of lutein and violaxanthine cycle pigments (Abadia et al. 1991; Quilez et al. 1992). This leads to an increase in the ratio of the lutein:chl a and violaxanthin:chl a in the leaves of iron-deficient plants (Abadia, 1992).

There are several components of photosystem I, wherein iron is involved in electron transport (Hopkins, 1983; Sandman and Malkin, 1983). This makes photosystem I sensitive to iron deficiency. Leaves of iron-deficient plants show a decrease in ferredoxin (Alcaraz et al. 1985; Huang et al. 1992) and cytochrome content per unit area (Spiller and Terry, 1980). A study of iron deficiency effect on photosynthetic apparatus of *Chlamydomonas reinhardti* by Moseley et al. (2002) showed a disconnection of the LHC I antenna from photosystem I (PS I) even before the iron-deficient cells developed chlorosis. The authors attributed this to observed changes in physical properties of the PS I-K. On persistence of iron deficiency, the photosynthetic apparatus undergoes some adaptive changes, such as degradation of existing proteins coupled to synthesis of new proteins that leads to a new steady state with decreased stoichiometry of electron transport complexes. Iron deficiency induced changes in PS I activities are largely reversed on recovery from the deficiency (Terry, 1983; Pushnik and Miller, 1989). Severe deficiency of iron also impairs PS II and the impairment is irreversible, possibly because of a structural damage to the PS II reaction centre. Bertamini et al. (2002) examined the effect of iron deficiency on the photosynthetic efficiency and PS II membrane organization of field-grown grapevine (*Vitis vinefera* L. Pinot noir). The chlorotic leaves (lime-induced chlorosis) showed more severe decrease in PS II activity than PS I activity, which could be attributed to severe degradation of D1, 33, 28-25 and 23 kDa polypeptides. Bertamini et al. (2002) inferred that the lime-induced chlorosis in grapevine leaves was the visible manifestation of accelerated degradation of the PS II light harvesting chloroplastic proteins (LHCP).

Information on iron deficiency effects on the reactions of photosynthetic carbon reduction cycle is limited. Sharma and Sanwal (1992) showed a decrease in rate of CO_2 fixation in leaves of iron-deficient maize plants. Arulananthan et al. (1990) showed an impairment in RuBP regeneration under iron deficiency, and attributed it to possible inhibition of Ru5P kinase.

2.4.4 Organic Acids

Iron-deficient plants show increased accumulation of organic acids, particularly citric and malic acids. The effect has been observed in both roots (Landsberg, 1981; Miller et al. 1990) and leaves (Palmer et al. 1963; Bindra, 1980; Welkie and Miller, 1993) and has been shown to be more pronounced in Fe-efficient rather than Fe-inefficient cultivars (Miller et al. 1990). Fournier et al. (1992) compared the iron deficiency effect on the accumulation of organic acids in two sunflower lines RHA291 and HA89, known to differ in Fe-efficiency. When subjected to iron deficiency, the roots of Fe-efficient line RHA291 accumulated high concentrations of organic acids, particularly citric acid but the less Fe-efficient line HA89 did not show any such response. Accumulation of organic acids in response to iron deficiency is generally attributed to enhanced proton extrusion by roots of iron-deficient plants, which contribute to the alkalization of the cytoplasm and favour the dark fixation of CO_2 (Gueren et al. 1991; Rabotti et al. 1995). Lopez-Millon et al. (2001) described iron deficiency effect on the concentration of organic acids and certain enzymes involved in the metabolism of organic acids in sugarbeet leaves. It was shown that deprivation of iron supply led to increase in the total organic acid concentration; and supply of iron to the Fe-deficient plants restored the total organic acids to levels found in Fe-sufficient plants. Increase in enzyme activities due to iron deficiency was also reversed on resupplying iron (Lopez-Millan et al. 2001). In a recent study, Bisht et al. (2002a) reported an increase in free (titrable) acids in fruits of sand-cultured tomato plants exposed to iron deficiency, which could be largely overcome when iron supply was raised adequately at the time of initiation of flowering. Lopez-Millan et al. (2001) suggested that enhanced accumulation of organic acids under Fe-deficiency was caused by their enhanced transport from roots to leaves, as also due to increase in carbon fixation in the leaves.

2.4.5 Nucleic Acids, Proteins

Iron-deficient plants show a decrease in the number of polyribosomes (Lin and Stocking, 1978) and ribonucleic acid (Spiller et al. 1987; Bisht et al. 2002a). Spiller et al. (1987) showed that a decrease in the mRNA and rRNA concentration in iron-deficient plants was reversed on recovery of plants from iron deficiency. They suggested a high requirement of iron for chloroplastic mRNA and rRNA. Bisht et al. (2002a) showed a decrease in the total RNA content in leaves of iron-deficient tomato plants, which could be reversed when the level of iron supply was raised from deficient to sufficient. They suggested low RNA content in iron-deficient plants to be a consequence of iron inhibition of RNase activity.

There are several reports of decrease in the protein concentration in leaves of iron-deficient plants (Perur et al. 1961; Shetty and Miller, 1966; Price et al. 1972; Bisht et al. 2002a).

2.4.6 Nitrogen Fixation

The nodulating legumes need a high dose of iron because it is a constituent of the nitrogenase complex, ferredoxin and leghaemoglobin, which are all involved in nitrogen fixation (Udavardy and Day, 1997). Iron fertilization of nodulating legumes not only benefits dry matter production, but also nitrogen content of plants (O'Hara et al. 1988; Tang et al. 1990). Iron deficiency has been shown to inhibit nodulation and N_2 fixation in *Lupinus angustifolius* (Tang et al. 1990, 1992) and *Arachis hypogaea* (O'Hara et al. 1988; Tang et al. 1991).

CHAPTER 3

MANGANESE

3.1 GENERAL

Manganese, lying in the centre of the first transition series, exists in several oxidation states. Of the common oxidation states in plants—Mn^{2+}, Mn^{3+}, Mn^{4+} and Mn^{5+}—the most dominant and stable is manganous (Mn^{2+}). It is readily oxidized to the less stable manganic form (Mn^{3+}). Manganese forms strong bonds with oxygen containing species. It forms ligands with aminoacid residues of the D I protein of the PS II reaction centre and functions as a charge accumulator by undergoing changes in its oxidation states. As a constituent or an activator of enzymes, manganese plays a catalytic role in cellular metabolism including CO_2 fixation in C_4 and CAM plants and detoxification of oxygen-free radicals. Plant response to inadequate supply of manganese is suggestive of its role in reproductive development, including seed set and development.

3.2 UPTAKE AND TRANSPORT

Plants absorb manganese as Mn^{2+}. Uptake was suggested to involve facilitated diffusion. Polarization of the plasmalemma due to proton pumping (H^+ efflux) is thought to cause acidification of the external solution, leading to enhanced Mn-uptake. Reflux of H^+ into the cytosol depresses plasmalemma polarization and inhibits Mn-uptake (Yan et al. 1992). However, reports of cation competition in uptake and genotypic differences in manganese uptake and transport suggested possible involvement of PM-bound manganese transporter in plants. Recent researches in this direction have led to isolation and characterization of several genes encoding possible manganese transporter proteins belonging to different unrelated gene families (Hall and Williams, 2003). Axelsen and Palmgren (2001) presented evidences suggesting participation of a P-type ATPase in manganese transport. The *Arabidopsis* transporter ECA1, which belongs to the Ca^{2+}-ATPase subfamily, was shown to complement Ca^{2+} pump defective yeast mutants grown on a medium containing manganese. This was attributed to the ability of ECA1 to form a Mn^{2+} dependent phosphoprotein (Liang et al. 1997). Subsequently, ECA1 was shown to function as an ER-bound Ca^{2+}/Mn-pump, which, under conditions of calcium deficiency or manganese excess, supports growth and confers tolerance to excess manganese (Wu et al. 2002). Recently, Delhaize et al. (2003) have isolated a proton: Mn^{2+} antiporter Sh MTP1 from *Stylosanthes halmata*. The Sh MTP1 belongs to the cation diffusion facilitator family transporters. When expressed in yeast mutants and in *Arabidopsis,* Sh MTP1 confers tolerance to manganese. Delhaize et al. (2003) have suggested that located on the internal membrane of the cell organelles, Sh MTP1 confers tolerance to Mn^{2+} through internal sequestration. Another cation/H^+ antiporter CAX2 (Cation exchanger 2) has been suggested to be involved in low-affinity manganese transport across the tonoplast (Hirschi et al. 2000). When expressed ectopically in yeast, CAX2 suppressed the yeast mutants sensitive to high manganese concentrations. Localized in the vacuolar membranes, CAX2 is suggested to participate in accumulation of Mn^{2+} (Ca^{2+} and Zn^{2+}) (Hirschi et al. 2000). Direct transport measurements in *Arabidopsis* have provided further evidence to suggest that CAX2 functions as Mn^{2+}/H^+ antiporter (Shigakai et al. 2003). There are also reports of 1RT1 involvement in manganese transport. Korshunova et al. (1999) showed that the *Arabidopsis* IRT1 complements the manganese-uptake deficient mutant of yeast, suggesting its involvement in manganese transport.

Nramp family transporters have also been suggested to be involved in manganese transport. Thomine et al. (2000) cloned *Nramp3* and *Nramp4* genes from *Arabidopsis* and showed them to complement manganese (and iron) uptake defective yeast mutants. It has been recently shown that Nramp3 has multiple substrate specificity (Mn, Fe, Cd) and is expressed in vascular bundles of root, stem and leaves (Thomine et al. 2003). Nramp3 has been suggested to play a role in mobilizing manganese (and other substrates) from the vacuolar pool and in their long distance transport (Thomine et al. 2003).

Long-distance transport of manganese through xylem takes place either in the ionic form or as chelates (Tiffin, 1972). White (1981) showed that in soybean and tomato, root to shoot transport of manganese takes place in the form of manganese complexes of citric and malic acids. Evidence on phloem mobility of manganese is conflicting (Welch, 1995). Most findings suggest poor mobility of manganese in phloem (Wittwer and Teubner, 1959; Hocking, 1980; Pearson and Rengel, 1994). The relative translocation of manganese to the leguminous fruits through the xylem and phloem depends on the manganese status of plants. Under conditions of limited availability of manganese, xylem input of manganese into the fruit is relatively small as compared to its input through the phloem, but in plants that are adequately supplied with manganese, the condition is reversed (Pate and Hocking, 1978). Studies using radioactive ^{55}Mn show low rates of translocation from leaves to the developing grains in wheat (Pearson and Rengel, 1994, 1995b). The damaged seeds of legumes, as also low seed vigour and viability—observed when mother plants are subjected to manganese deficiency—have been suggested to be a consequence of limitation in Mn-translocation to the developing seeds (Crosbie et al. 1993; Longnecker et al. 1996).

3.3. ROLE IN PLANTS

3.3.1 Enzyme Action

Manganese plays a catalytic role as a constituent or activator of over 30 enzymes (Burnell, 1988). While requirement for manganese is specific for some enzymes, many enzymes activated by magnesium (Mg^{2+}) are also activated by manganese and by other dimetal cations.

Mn-Superoxide Dismutase (EC. 1.15.1.1)

The mitochondrial superoxide dismutase (Mn-SOD) is a manganese metalloprotein. It is a dimeric protein with ~ 20 kDa subunits. Each subunit contains a single manganese ion (Mn^{3+}) which binds to three imidazole (histidine) groups, a carboxyl group and a hydroxyl ion ($^-$OH) or H_2O with a site open for binding to the substrate (O_2^-). The enzyme catalyzes the dismutation of two superoxide ions (O_2^-) to a molecule of hydrogen peroxide, accompanied by generation of a molecule of oxygen. During the process, two protons are taken up and manganese undergoes oxidation-reduction ($Mn^{3+} \leftrightarrows Mn^{2+}$).

$$2O_2^- + 2H^+ \longrightarrow H_2O_2 + O_2$$

Mn-SOD functions as a component of the antioxidative system of the mitochondria, which forms major site for generation of reactive oxygen species.

Phosphoenol pyruvate carboxykinase (EC 4.1.1.49)

Phospho*enol* pyruvate carboxykinase (PEPCK), which catalyzes the reversible conversion of oxaloacetate to phospho*enol*pyruvate, has an absolute requirement of manganese (Mn^{2+}) that cannot be substituted by Mg^{2+}. PEPCK is of particular importance to C_4 (aspartate type) plants, where it catalyzes the decarboxylation of oxaloacetate to phospho*enol*pyruvate in the bundle sheath chloroplasts. The CO_2 thus released serves as the entry point for the Calvin cycle

$$\text{Oxaloacetate} + ATP \rightleftharpoons \text{Phospho}enol\text{pyruvate} + ADP + CO_2$$

Recently, Chen et al. (2002b) have shown that PEPCK of guinea grass (C_4) *Panicum maximum* has a high affinity for CO_2 and suggested that it may catalyze the reverse reaction by functioning as a carboxylase.

NAD⁺ - MALIC ENZYME (EC 1.1.1.39)

The NAD^+ - Malic Enzyme (Malate - NAD^+ oxidoreductase, decarboxylating) of C_4 (aspartate) plants specifically requires manganese (Mn^{2+}) for activation (Hatch and Kagawa, 1974). The enzyme catalyzes the oxidative decarboxylation of malate to pyruvate in the mesophyll cells of C_4 plants.

$$\text{Malate} + NAD^+ + H^+ \rightleftharpoons \text{Pyruvate} + CO_2 + NADH$$

The CO_2 generated in the reaction serves as a substrate for Rubisco in the bundle sheath chloroplasts and provides the point of entry into the Calvin cycle.

ALLANTOATE AMIDOHYDROLASE (EC 3.5.3.9.)

The manganese-containing allantoate amidohydrolases, which catalyze the oxidative decarboxylation of allantoate to uredoglycine, require manganese as a cofactor (Winkler et al. 1987; Lukaszewski et al. 1992).

$$\text{Allantoate} \xrightarrow{\quad CO_2 \quad} \text{Uredoglycine} + NH_3 + CO_2$$

The enzyme is involved in the cataboism of ureides in the group of legumes (including soybean, cowpea) wherein nitrogen fixed in the root nodules is exported to the aerial part as ureides (allantoate, allantoate).

NADP⁺ - MALATE ENZYME (EC 1.1.1.40)

$NADP^+$-Malate dehydrogenase, which catalyzes the oxidative decarboxylation of malate to generate pyruvate and NADPH in C_4 (malate type) plants, is activated by magnesium as well as manganese.

$$\text{Malate} + NADP^+ + H^+ \rightleftharpoons \text{Pyruvate} + CO_2 + NADPH$$

The $NADP^+$ - malate dehydrogenase functions in C_4 photosynthesis. It is also important for fatty acid biosynthesis in lipid storing seeds (castor seed endosperm), wherein oxidative decarboxylation of pyruvate to acetyl CoA provides the carbon skeleton for fatty acid biosynthesis (Smith et al. 1992).

ISOCITRATE DEHYDROGENASE (EC 1.1.1.42)

Isocitric dehydrogenase (IDH) catalyzes pyridine nucleotide-dependent conversion of isocitrate to α-ketoglutarate. The enzyme is activated by magnesium as well as manganese. The conversion of isocitrate to α-ketoglutarate is catalyzed in two steps. The first step involves the oxidation of isocitrate to oxalosuccinate. In the second step, the unstable oxalosuccinate molecule is decarboxylated to α-ketoglutarate.

$$\text{Isocitrate} \longrightarrow \text{Oxalosuccinate} \xrightarrow{\quad CO_2 \quad} \text{α-ketoglutarate}$$

The plant enzyme exists in two forms which differ in their localization and requirement of pyridine nucleotide as a co-substrate. The mitochondrial isoform has a specific requirement of NAD^+ (NAD^+-Isocitrate dehydrogenase, EC 1.1.1.41). It functions as a component of the citric acid cycle and provides the first molecule of NADH during the turn of the cycle. The plant enzyme differs from its mammalian counterpart in the sense that ATP does not inhibit it. Instead, it is inhibited by NADH and is very sensitive to NAD: NADH ratios. The other form is found in the cytosol, plastids, peroxisomes and also mitochondria, and is $NADP^+$ dependent ($NADP^+$-Isocitrate dehydrogenase, EC 1.1.1.42). The cytosolic IDH is the main source of carbon skeleton for the assimilation of ammonium. It provides a continuous supply of α-ketoglutarate in such quantities as are required for the assimilation of NH_4^+ in into amino acids involving glutamine synthetase and glutamate synthase (GOGAT). The cytosolic enzyme not only contributes to amino acid synthesis but also prevents the accumulation of NH_4^+ in toxic concentrations.

PHOSPHO*ENOL*PYRUVATE CARBOXYLASE (EC 4.1.1.31)

Phospho*enol*pyruvate carboxylase (PEPCase), essentially a Mg^{2+} enzyme, is also activated by Mn^{2+}. The spinach PEPCase contains 12 subunits, each containing a bound manganese ion (Mn^{2+}). The enzyme catalyzes the carboxylation of phospho*enol*pyruvate to form oxaloacetate, with carboxyl phosphate as an unstable intermediate.

$$\text{Phospho}enol\text{ pyruvate} + HCO_3 \xrightarrow{\quad Pi \quad} \text{Oxaloacetate}$$

The enzyme plays a key biosynthetic role in plants. It is involved in 'priming' the citric acid cycle and biosynthesis of amino acids (aspartate, glutamate). In C_4 plants, the enzyme functions as a CO_2 concentrating mechanism.

GLUTAMINE SYNTHETASE (EC 6.3.1.2)

Glutamine synthetase (GS), involved in the primary pathway of ammonium metabolism, requires Mg^{2+} or Mn^{2+} for activation. The enzyme catalyzes the amidation of glutamate to glutamine in an ATP-dependent reaction.

$$\text{Glutamate} + NH_4^+ + ATP \longrightarrow \text{Glutamine} + ADP + Pi$$

Glutamine synthetase plays an important role in the regulation of nitrogen metabolism.

ENOLASE (EC 4.2.1.11)

Manganese (Mn^{2+}), as also magnesium (Mg^{2+}), activates enolase, an enzyme involved in the reversible dehydration of 2-phosphogycerate to form phospho*enol*-pyruvate (PEP), the phospho-derivative of the *enol* form of pyruvic acid.

$$\text{2-Phosphoglycerate} \rightleftharpoons \text{Phospho}enol\text{pyruvate} + H_2O$$

The dehydration of phosphoglycerate markedly elevates the transfer potential of the phosphoryl (*enol* phosphate) group, allowing easy transfer of its phosphate group to ADP, leading to the formation of ATP by pyruvate kinase. The enzyme plays a role in sugar metabolism and is very responsive to environmental stresses such as high temperature. It has been suggested to act as a general stress protein involved in protection of cellular components (Schweisguth et al. 1995).

3.3.2 Photosynthesis

Early work on the effect of manganese deficiency on photosynthetic O_2 evolution (referred under 3.4.3) led to investigations that substantiated a role of manganese in oxidation of water in PS II (Ono and Onone, 1991; Renger and Wydrzynski, 1991; Hoganson and Babcock, 1997). More conclusive evidence of this came from flash photolysis studies using Mn-depleted PS II (Hoganson et al. 1993). Investigations using x-ray absorption spectroscopy (Yachandra et al. 1993) led to development of a structural model showing that the PS II complex contained four (two pairs) manganese atoms in a cluster linked to (Fig.3.1) into a proximal calcium

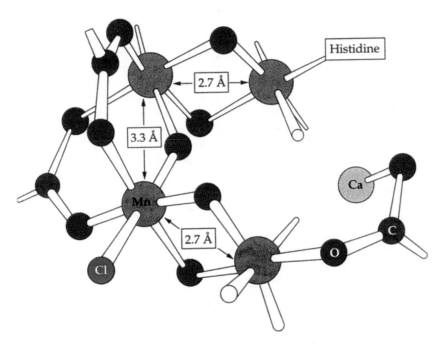

Fig. 3.1. Model of the Mn-cluster of the PS-II complex. Four Mn atoms are liganded to the aminoacids residue of PS-II protein D1, as well as to oxygen and chloride. (Reprinted, with permission, from *Biochemistry and Molecular Biology of Plants.* BB Buchnam, W Gruissem, RL Jones, eds ©2000 American Society of Plant Biologists.).

atom and a halide (Yachandra et al. 1993; Ananyev and Dismukes, 1997). The manganese cluster binds to the aminoacid residues of the DI protein of PS II core complex (Campbell et al. 2000; Ono, 2001) and functions as a locus for charge accumulation. The manganese cluster passes through a series of oxidation states (S_0 to S_5) and provides a binding site for the water molecules and accumulation of the charge generated by four successive quanta absorbed by P680. Through this process, four electrons are extracted from water and a molecule of oxygen is released.

Being an activator of enzymes involved in the carboxylation and decarboxylation of C_4 acids, manganese plays an important role in C_4 photosynthesis. The first reaction of C_4 photosynthesis involves carboxylation of phospho*enol* pyruvate (PEP) to oxaloacetate (C_4 organic acid) by PEP carboxylase, which is activated by Mn^{2+}. In both malate and aspartate type plants, Mn^{2+} is involved in decarboxylation of C_4 acids in the bundle sheath cells to produce CO_2, which is refixed in a C_3 compound by Rubisco. While $NADP^+$ malic enzyme, that catalyzes decarboxylative dehydrogenation of malic acid in malate type plants, is activated by both Mn^{2+} and Mg^{2+}, NAD^+ malic enzyme and PEP carboxykinase, catalyzing decarboxylation of C_4 acids in the two variants of the aspartate type plants, have a specific requirement of Mn^{2+} for activation (Burnell, 1986). In this way, manganese contributes to a CO_2 concentration mechanism in the bundle sheath cells of C_4 plants that favour carboxylase activity of Rubisco to its oxygenase activity.

3.3.3 Secondary Metabolism

Biosynthesis of secondary metabolites in plants involves multistep pathways, some steps being catalyzed by manganese-activated enzymes (Burnell, 1988; Hughes and Williams, 1988). Synthesis of isopentenyl diphosphate (IPP) in cytosol and ER, via acetate-mevalonate pathway involves conversion of mevalonic acid to mevalonic acid-5-phosphate, catalyzed by the manganese activated mevalonic acid kinase (Lalitha et al. 1985).

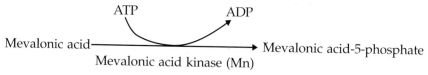

Isopentenyl diphosphate (IPP) serves as the precursor for biosynthesis of terpenoids and provides branch points for the synthesis of several other secondary products such as the chloroplast pigments (phytol, carotenoids) and gibberellins (Fig. 3.2). Two steps in the conversion of IPP (C_5), synthesized in plastids (via pyruvate-glyceraldehyde 3-phosphate pathway) to phytoene (C_{40}) are catalyzed by Mn^{2+} activated enzymes. The conversion of dimethyl allyl diphosphate to geranyl diphosphate is catalyzed by geranyl diphosphate synthase (Croteau and Purkett, 1989) and that of geranyl diphosphate to phytoene by phytoene synthase (Maudinas, 1977), both of which require manganese for activation.

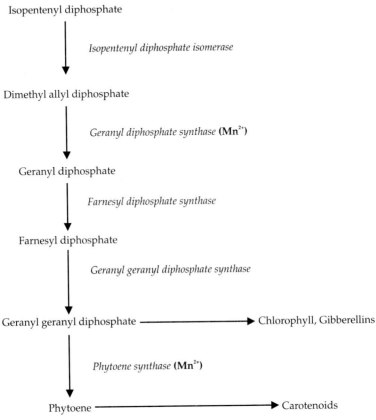

Fig. 3.2. Scheme for carotenoid biosynthesis showing steps catalyzed by Mn^{2+} - activated enzymes

Geranyl geranyl diphosphate (GGPP) also provides the starting point for the biosynthesis of phytol sidechain of chlorophylls and gibberellins. On way to gibberellin biosynthesis, GGPP undergoes a two-step conversion to *ent* – kaurene. Synthesis of *ent* kaurene, the first committed precursor for gibberellins is catalyzed by kaurene synthase which requires Mn^{2+}, (Co^{2+} or Mg^{2+}) for activation.

56

Manganese is also involved in the biosynthesis of aromatic amino acids—phenylalanine, tyrosine and tryptophan—that function as precursors of a diverse group of compounds such as flavanoids, indole, and lignin (Burnell, 1988; Hughes and Williams, 1988). Phenylalanine and

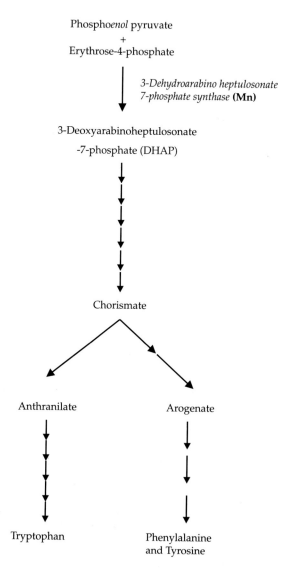

Fig 3.3. Scheme for biosynthesis of aromatic amino acids in plants, the first reaction of which involves catalysis by a Mn-activated enzyme

tyrosine, which form the starting material for phenylpropanoid metabolites, including phenols, phenolic alcohols and their polymerization products (lignans, lignins), are synthesized through the shikimate-chorismate-arogenate pathway (Fig. 3.3). The first reaction of this pathway involves condensation of phosphoenol pyurvate with erythrose-4-phosphate to produce 3-Deoxyarabinoheptulosonate-4-phosphate (DAHP) catalyzed by DAHP synthase, the plastidic form of which is activated by Mn^{2+} (Hermann, 1995; Hermann and Weaver, 1999).

Being an activator of arginase (Burnell, 1988), manganese plays a role in the synthesis of polyamines, which play important roles in plant growth and development, and also in detoxification of active oxygen species (Evans and Malmberg, 1989).

3.4 DEFICIENCY RESPONSES

3.4.1 Visible Symptoms

Manganese deficiency generally manifests itself as small chlorotic patches in the interveinal areas of the subterminal or upper middle leaves. Gradually, the patches increase in number and size and coalesce to form chlorotic stripes in case of parallel veined leaves (as in rice, barley) and fine chlorotic mottling in reticulately veined leaves of dicotyledonous plants. In case of prolonged or severe deficiency, the interveinal areas turn necrotic and develop buff, brown or reddish-brown colouration, often followed by splitting of the lamina. In early description of manganese deficiency in sugarcane, the more or less countinuous red stripes followed by splitting of the lamina along the stripes were referred as 'Pahala blight'. Sugarbeet and mangold are particularly sensitive to manganese deficiency. Leaves of manganese-deficient plants of sugarbeet show yellow-green mottling followed by chlorosis, necrosis and eventually breakdown of lamina. The leaf margins roll inward giving the leaves a sagitate or an arrow-shaped appearance and the leaves tend to remain more upright than in normal plants. Amongst cereals, oats are very sensitive to manganese deficiency. Middle and old leaves of manganese-deficient oat plants develop grey or buff-coloured patches, some distance away from the base; the patches are more numerous near the margins and appear more conspicuous on the undersurface of leaves. These patches gradually enlarge and coalesce to form more or less continuous streaks, which eventually turn brown. Somewhat similar symptoms are produced in rice, another cereal sensitive to manganese deficiency. The interveinal chlorotic patches produced in the middle of the leaf blades turn reddish brown (Agarwala and Sharma, 1979). The leaf sheath and the stem internodes may also develop similar symptoms. If severe deficiency persists, the entire plants give a bronzed appearance. Tillering, panicle emergence and grain filling is severely inhibited and a large percentage of glumes remain empty (Longnecker et al 1991a).

Under moderate deficiency of manganese, which permits plants to go reproductive, seeds of beans and peas show browning, necrosis and eventually disintegration of the central part of the cotyledons. In early observations, such an effect in peas was described as 'marsh spot'.

3.4.2 Enzyme Activities

Deprivation of manganese supply leads to changes in the activities of enzymes that contain manganese as a cofactor or require manganese for activation. Manganese-deficient plants show low activity of malic dehydrogenase (Agarwala et al. 1986; Sharma et al. 1986). In rice, the decrease in malic dehydrogenase activity under manganese deficiency has been shown to be more in plants-supplied with nitrogen as ammonium nitrate than those supplied with nitrate nitrogen. Manganese deficiency alters the activity of superoxide dismutase but the effect is not consistent and also varies with the metal cofactor of the enzyme. Del Rio et al. (1978b), Leidi et al. (1987b) and Kröniger et al. (1995) observed a decrease in Mn-SOD activity under manganese deficiency. On the contrary, Mehlorn and Wenzel (1996) and Yu and Rengel (1999) found enhanced activity of Mn-SOD in manganese-deficient plants. Overexpression of Mn-SOD in transgenic tobacco plants has been shown to enhance tolerance to freezing injury (Mc Kersie et al. 1993) and oxidative damage (Slooten et al. 1995). Yu et al. (1998) observed an increase in the activity of Cu-Zn SOD in response to manganese deficiency and attributed it to manganese deficiency induced increase in accumulation of Cu and Zn.

Manganese deficiency is reported to decrease the activity of allantoate amidohydrolase (EC 3.5.3.9), leading to enhanced accumulation of ureide. Increase in manganese supply increases the rate of ureide degradation and N_2 fixation in soybean under conditions of water deficit (Purcell et al. 2000; Vadaz et al. 2000). Not all cases of enhanced accumulation of ureide can, however, be attributed to impairment in the activity or to the role of manganese as an activator of allantoate amidohydrolase (Vadez and Sinclair, 2002). Oxidative decarboxylation of allantoate is also catalyzed by another enzyme, allantoate amidinohydrolase (EC 3.5.3.4) that does not involve a manganese cofactor (Shelp and Ireland, 1985). In soybean cv. Jackson, ureide degradation does not involve manganese as a cofactor. Instead, it is catalyzed by allantoate amidinohydrolase. Sinclair et al. (2003) did not find ureide accumulation to be related to manganese in six out of the eight soybean lines with N_2 fixation tolerance to soil drying.

Activities of several other enzymes that have no specific manganese requirement, undergo changes in response to manganese deficiency (Khurana et al. 1991, 1999; Chatterjee et al. 1989, 1994). Manganese-deficient plants show decreased activity of acid phosphatase (Khurana et al. 1991) and ATPase (Chatterjee et al. 1994). Agarwala et al. (1986) showed that manganese effect on activities of iron enzymes, aconitase and succinic dehydrogenase in rice is influenced by the form of nitrogen supply, which influences the response of rice to iron deficiency. In plants-supplied nitrogen as ammonium nitrate, activities of aconitase and succinic dehydrogenase showed little change in response to manganese deprivation, but in plants-supplied nitrogen as nitrate, activities of these enzymes showed significant increase in response to manganese deficiency. This may possibly involve influence of nitrogen source on the physiological availability of iron. Manganese effect on the two haem enzymes, catalase and peroxidase, is not identical. Almost invariably, low manganese plants show increased activity of peroxidase with little effect or decrease in the activity of catalase (Agarwala et al. 1964; Bar-Akiva and Lavon, 1967; Chatterjee et al. 1989, 1994). Mn deficiency is also reported to cause enhancement in the activity of IAA oxidase (Morgan et al. 1976). Manganese is also reported to regulate the activity of nitrate reductase (Leidi and Gomez, 1985). Diverse results have been reported in respect of manganese-deficiency effect on ribonuclease (RNase) activity. While Sharma et al. (1986) reported enhanced activity in the leaves of manganese-deficient sugarcane, Chatterjee et al. (1994), reported a decrease in RNase activity in leaves of manganese-deficient wheat plants.

3.4.3 Photosynthesis

Much before manganese was shown to be a constituent of the PS II reaction centre, algal cultures deprived of manganese were reported to show a decrease in photosynthetic O_2 evolution (Pirson, 1937, Kessler et al. 1957; Eyster et al. 1958). Similar observations were later extended to higher plants. There are several reports of Mn-deficiency induced decrease in the rate of photosynthesis (Cheniae and Martin, 1970; Terry and Ulrich, 1971; Nable et al. 1984; Kriedemann et al. 1985; Debus, 1992). Mn-deficiency effects on photosynthesis have also been shown to be reversed on resumption of the Mn-supply (Nable et al. 1984; Kriedemann and Anderson, 1988). Inhibition of photosynthesis under Mn-deficiency is essentially attributed to its role as a constituent of PS II (Hoganson et al. 1993; Yachandra et al. 1993). Inhibition of photosynthesis could also result from manganese deficiency induced changes in leaf ultrastructure (Possingham et al. 1964; Polle et al. 1992). In Norway spruce, Mn-deficiency has been shown to cause enhanced peroxidation of the membrane lipids and structural damage to the thylakoids (Polle et al. 1992). In peppermint, manganese deficiency is reported to cause a significant decrease in CO_2 exchange rates, chlorophyll content and essential oil content (Srivastava and Luthra, 1994b) which were reversed on resupply of manganese. Srivastava and Luthra (1994) inferred that manganese deficiency reduces the availability of photoassimilates for essential oil biosynthesis because of their preferential utilization in some primary metabolic processes.

3.4.4 Carbohydrates, Fatty Acids

Manganese deficiency is reported to induce changes in concentration of carbohydrates and fatty acids. In general, plants exposed to manganese deficiency display enhanced accumulation of soluble carbohydrates. There are several reports of decrease in fatty oil content of seeds of manganese-deficient plants (Heenam and Campbell, 1980, Wilson et al. 1982; Campbell and Nable, 1988; Sharma et al. 1995).

3.4.5 Nitrogen Metabolism, Nucleic Acids

Manganese-deficient plants show enhanced accumulation of soluble nitrogenous compounds. Because of manganese involvement in activation of arginase (Burnell, 1988) and allantoate amidohydrolase (Winkler et al. 1985), Mn deficiency leads to increased accumulation of arginine and ureides. Manganese-deficient plants of sugarcane are reported to be low in nucleic acids (Sharma et al. 1986; Chatterjee et al. 1994). Decrease in RNA could either result from decrease in its synthesis, catalyzed by RNA polymerase which requires Mn^{2+} for activation (Ness and Woolhouse, 1980), or from its enhanced degradation caused by elevated levels of ribonuclease under manganese deficiency (Sharma et al. 1986).

3.4.6 Secondary Metabolism

Concentration of several secondary metabolites of terpenoid biosynthesis pathway and shikimate-chorismate-arogenate pathway is reported to be influenced by manganese nutrition of plants. Wheat plants deprived of manganese supply showed decreased concentration of chloroplast pigments (Wilkinson and Ohki, 1988). Brown et al. (1984) reported manganese deficiency induced decrease in the concentration of phenolics and lignin, which could be attributed to change in activities of manganese-activated enzymes. Manganese-deficient plants show decreased activity of phenylalanine ammonia lyase, which catalyzes the conversion of phenylalanine to cinnamic acid, which is subsequently converted to cinnamic alcohols and finally to lignin.

3.4.7 Defense Against Oxidative Damage, Pathogenic Diseases

As a cofactor of the mitochondrial superoxide dismutase (Mn-SOD), a constituent enzyme of the plants' antioxidant system, manganese provides protection against oxidative damage. By catalyzing dismutation of O_2^- to H_2O_2, it prevents the damage resulting from excessive accumulation of superoxide ions and their derivatives in the mitochondria (Bowler et al.

1991). Inadequate supply of manganese may enhance plant disposition to oxidative stress. Transgenic tobacco overexpressing Mn-SOD in the chloroplasts is reported to show enhanced tolerance to oxidative stress (Slooten et al. 1995). Mn-SOD is also suggested to contribute to drought tolerance in plants. Wu et al. (1999) described an increase in the Mn-SOD expression in wheat seedlings on the induction of drought-like conditions and a decrease in its expression on rehydration of the seedlings.

Manganese provides resistance against pathogenic diseases (Graham, 1983; Huber and Wilhelm, 1988; Huber and Graham, 1999). The incidence of take all disease of wheat in Western Australia has been shown to be related to manganese concentrations in plants (Brennan, 1992). Low manganese status of plants rendered them susceptible to the disease. Rengel et al. (1994) found differences in wheat genotypes to take all to be related to the difference in the pattern of biosynthesis of phenolics and lignin in their roots and suggested that manganese involvement in the biosynthesis of such compounds, which inhibits the penetration of the pathogens into the host cells, contributes to its protective role against pathogenic fungi. Manganese fertilization is reported to decrease the incidence of take all in wheat (Graham and Rovira, 1984; Huber and Wilhelm, 1988), brown spot of rice (Kaur et al. 1979) and rust in soybean (Morab et al. 2003). Foliar sprays of 0.3% $MnSO_4$ to the rust-susceptible soybean variety JS-335 has been reported to decrease the per cent disease index of soybean following inoculation with *Phakospora pachyrrhizi* Syd. from 90% (untreated control) to 34%.

3.4.7 Reproductive Development and Yield

Manganese effect on yield attributes is a major factor contributing to low yield of manganese-deficient plants. In barley, manganese deficiency inhibits tillering and thus the number of effective (ear bearing) tillers (Longnecker, 1991a). In cotton, wherein the flowers are borne on the tertiary branches, capsule yield is severely limited by Mn deficiency because of inhibition of tertiary branching (Pandey et al. 2002a). Sharma (1992) has shown that, for the same level of deficiency, reduction in the reproductive yield of wheat is more marked than vegetative yield. This is ascribed to critical requirement of manganese for reproductive development. In wheat, manganese deficiency limits grain yield not only by inhibiting tillering and the number of ears plant[-1] but also by causing pollen-sterility and reducing the number of seeds ear[-1]. Manganese deficiency has been reported to decrease the pollen-producing capacity, size of the pollen grains, their cytoplasmic contents pollen tube growth (Sharma et al. 1991, Sharma, 1992) and activities of some anther enzymes (Kaur et al. 1988).

Hand pollination studies by Sharma et al. (1991) showed that when emasculated ears of the manganese-sufficient plants were pollinated with pollen from manganese-deficient plants, the number of seeds ear[-1] was reduced much below that in Mn-sufficient plants. Cross fertilization of manganese-deficient plants with manganese-sufficient pollen produced about the same number of seeds ear[-1] but lower seed weight than in manganese-sufficient plants (Table 3.1). It was inferred that manganese deficiency retards both fertilization and seed maturation. Impairment in manganese supply affects seed quality in pea (Khurana et al. 1999). Manganese-deficient plants of lupin show discolouration, splitting and deformity of seeds (Campbell and Nable, 1986).

Table 3.1. Manganese effect on seed set by cross-pollination in maize (*Zea mays* L cv. **G2**).

♂	Cross X	♀	Cob weight (g)	Seed weight (g cob[-1])	Seeds set (No. cob[-1])	Seed weight (mg)
+Mn		+Mn	142	108.0	353	305
-Mn		+Mn	37	8.8	23	382
-Mn		-Mn	34	9.1	25	364
+Mn		-Mn	86	65.1	287	31

(After Sharma et al. 1991)

Analysis of manganese levels at different stages of seed development in *Arabidopsis* (Otegui et al. 2002) shows storage of Mn (as Mn-phytate) in the endoplasmic reticulum of the chalazal region of the endosperm, and its mobilization to the embryo at the late bent-cotyledon stage. The disappearance of Mn- phytate from the endosperm coincides with the accumulation of two Mn-binding proteins of the embryo—the Mn-SOD and the 33KD protein of the Mn-complex of PS II. It is suggested that Mn stored in the endosperm as Mn-phytate functions in the assembly of Mn-proteins during embryo development (Otegui et al. 2002).

CHAPTER 4

COPPER

4.1 GENERAL

Copper is a redox-active transition metal. It exists in two oxidation states, cuprous (Cu^+) and cupric (Cu^{2+}). The more stable form is cuprous (Cu^+). The ability of copper ions to undergo reversible change in oxidation states ($Cu^+ \rightleftharpoons Cu^{2+}$) enable copper to function in cellular oxidation-reduction processes. Like iron, copper provides sites for reaction with molecular oxygen. It has great affinity to combine with organic ligands. Many copper proteins function as enzymes and electron carriers catalyzing oxidation-reduction reactions in cellular metabolism. As a constituent of cytochrome c oxidase, copper functions as a terminal component of the mitochondrial electron transport system. The blue copper protein *plastocyanin* functions as single electron carrier during photosynthetic electron transport. A similar copper protein *phytocyanin* has been reported from cucumber seeds (Guss et al. 1996). Copper is also involved in the detoxification of superoxide radicals, lignification of plant cell walls and in pollen fertility.

4.2 UPTAKE, TRANSPORT, DISTRIBUTION

Plants take up copper as Cu^+, to which state it is reduced by plasmalemma-bound cupric reductases (Hasset and Kosman, 1995; Georgatsou et al. 1997). This possibly involves the same plasma membrane reductase system, which carries out the reduction of Fe^{3+} (Welch et al. 1993; Cohen et al. 1997). Cohen et al. (1997) have shown enhanced induction of root cell plasma membrane ferric reductase and acidification of the rooting medium in response to copper deficiency. Romheld (1991) has suggested a role of phytosiderophores in copper uptake by graminaceous plants.

Copper uptake is an active process, with specific transporter proteins mediating transport across the plasmalemma (Fox and Guerinot, 1998). Kampfenkel et al. (1995) isolated a putative copper transporter COPT1 which showed high sequence similarity to yeast gene *CTR2* that encodes a low-efficiency copper transporter, and complements the copper-uptake defect in yeast mutants. *COPT1* is reported to be expressed in leaves, stem and flowers but not in roots (Kampfenkel et al. 1995). Recently Sanceñon et al. (2003) have identified five putative copper transporters COPT1, COPT2, COPT3, COPT4 and COPT5 from *Arabidopsis*. The *COPT 1-5* transcripts were detected in leaf, stem, flowers and also roots but their abundance differed in different plant parts. *COPT1, 2, 3,* and *5,* more so *COPT1,* and *COPT2,* functionally complemented high-affinity copper transport yeast mutant and were downregulated under copper excess (Sanceñon et al. 2003).

Long-distance transport of Cu^{2+} takes place in the form of copper complex with some amino acids (Loneragan, 1981). In soybean, copper transport mostly takes place as a complex with asparagine and histidine and in tomato as a complex with histidine, asparagine and glutamic acid (White et al. 1981; Loneragan, 1981). Both xylem and phloem mobility of copper is low (Hocking, 1980) and influenced by copper status of plants. Movement of copper from leaves to the developing grains is severely limited under copper deficiency (Loneragan, 1981). In the common vetch, during the pod filling stage, very little copper is translocated from the leaves into the seeds (Caballero et al. 1996). Fox and Guerinot (1998) have suggested the possibility of functioning of multiple copper trafficking pathways for delivery of copper to its different functional sites. Several intracellular metal transport proteins, known as metallochaperones, have been shown to be involved in the intracellular transport of copper. The copper–chaperone proteins are involved in the transport of Cu^+ ions from the cytoplasm to the functional sites at which it is incorporated into the functional proteins, e.g. Cu-Zn SOD (Huffman and O' Halloran, 2001).

4.3 ROLE IN PLANTS

4.3.1. Enzyme action

Copper is cofactor of a number of plant oxidases including mixed function oxygenases, that cause reduction of one atom of molecular oxygen (O_2) to water and utilize the other atom for oxygenation or hydroxylation of substrates (e.g. the phenolase complex). The active centre of the copper enzymes contains the metal in different oxidation states - Cu^{2+} (type 1 and type 2), Cu^+ (type 3) or both. Copper is also involved as a cofactor in some multimetal enzymes, such as superoxide dismutase and cytochrome c oxidase.

In the recent past, the incorporation of copper ions into enzyme proteins such as cytochrome c oxidase and Cu-Zn superoxide dismutase, has been a subject of active investigations. Using the yeast model, a new family of soluble metal receptor proteins, termed copper *metallochaperones*, have been shown to function in the delivery of copper to the Cu-proteins. They function in a chaperone like manner, guiding the copper ion to the target proteins, offering protection from other copper-binding sites (Lippard, 1999; Rae et al. 1999; O'Halloran and Culotta, 2000).

ASCORBATE OXIDASE (EC 1.10.3.3)

Ascorbate oxidase is a dimeric blue copper protein. Eight copper ions (Cu^{2+}) are bound mol^{-1} enzyme proteins. The enzyme catalyzes the dehydrogenation of ascorbate (the ionized form of ascorbic acid) to dehydroascorbate, coupled to the four-electron reduction of molecular oxygen to water:

$$2L \text{ Ascorbate} + O_2 \rightarrow 2\text{Dehydroascorbate} + 2H_2O$$

The reaction involves a valency change ($Cu^{2+} \rightleftharpoons Cu^{+}$), and is catalyzed in two successive one-electron reactions involving mono-dehydroascorbate as an intermediate. The enzyme is of wide occurrence in cell walls and cytoplasm of plant cells.

CATECHOL OXIDASE (EC 1.10.3.1)

Catechol oxidase, typifying polyphenol oxidases or phenolases, is a copper protein. It contains four copper ions mol^{-1}. The enzyme catalyzes the oxidation of o - or p - diphenols to quinones coupled to the reduction of molecular oxygen to water.

$$2O\text{-diphenol} + O_2 \longrightarrow 2O\text{-quinone} + 2H_2O$$

Polyphenol oxidases are important for secondary metabolism of plants. They are involved in flavonoid metabolism and biosynthesis of lignin and alkaloids. They also account for the formation of melanotic substances produced on the cut (wounded) surfaces of fruits and of phytoalexins that inhibit fungal growth.

LACCASE (EC 1.10.3.2)

Laccase (p-diphenol oxidase), first isolated from latex of Japanese lac tree, is a copper enzyme. It contains four copper ions of mixed valency states ($Cu^{2+} Cu^{+}$) per mol^{-1} and catalyzes the oxidation of p-diphenols to p-quinones. Catalysis involves a free radical mechanism, coupled with the reduction of molecular oxygen to water.

$$2p\text{-diphenol} + O_2 \longrightarrow 2p\text{-quinone} + 2H_2O$$

Laccase can also catalyze coupling reactions of monolignols (using O_2 instead of H_2O_2 as the electron acceptor), leading to lignin biosynthesis. It is also possibly involved in the synthesis of chloroplastic quinone (plastoquinone) involved in PS II electron transport (Ayala et al. 1992).

TYROSINASE (EC 1.14.18.1)

First identified in mushrooms, but widely distributed in plants, tyrosinase typifies a mixed function monophenol monoxygenase, in which one oxygen atom from molecular oxygen is incorporated in the substrate (tyrosine), and the other is reduced to water. The enzyme contains four copper ions mol^{-1} and is involved in the conversion of tyrosine (or o-monophenol) to dopa-quinone in a two-step reaction. In the first reaction (*cresolase*), the enzyme catalyzes the hydroxylation of tyrosine to 3,4 dihydroxyphenylalanine (DOPA). The second reaction (*catecholase*) involves the oxidation of dopa-quinone, coupled with the reduction of molecular oxygen to water. The overall reaction can be summarized as:

$$\text{Tyrosine} + \text{dihydroxyphenylalanine} + O_2 \longrightarrow$$
$$\text{Dihydroxyphenylalanine} + \text{DOPA-quinone} + H_2O$$

Unlike the animal tyrosinase, which is relatively specific for L-tyrosine and dihydroxy-L-phenylalanine (dopa), the plant enzyme acts upon a large variety of monophenols.

DIAMINE OXIDASE (1.4.3.6.)

Copper is an intrinsic constituent of certain diamine oxidases (DAO). They catalyze the oxidative deamination of several biologically active amines (e.g. putrescine, cadaverine) with the production of the corresponding aminoaldehydes, ammonia and hydrogen peroxide.

$$RCH_2NH_2 + O_2 + H_2O \rightarrow RCHO + NH_4 + H_2O_2$$

Copper amine oxidases are of common occurrence in leguminous plants during early stages of growth (Walker and Webb, 1981; Laurenzi et al. 2001). They are suggested to play a role in regulating the H_2O_2 supply for peroxidase activity involved in lignin/suberin synthesis. Since H_2O_2 is cytotoxic to pathogens and activates the defence genes (Lamb and Dixon, 1997), enhanced production of H_2O_2 resulting from activation of Cu amine oxidase may contribute to defence against pathogens (Rea et al. 2002).

SUPEROXIDE DISMUTASE (EC 1.15.1.1)

The chloroplastic superoxide dismutase (Cu-Zn SOD) is a homodimer, the active site of which contains one atom each of copper (Cu^{2+}) and zinc (Zn^{2+}) in each subunit. The enzyme carries out the disproportionation of the superoxide ions (O$_2^-$) to hydrogen peroxide (H$_2$O$_2$) and oxygen (O$_2$).

$$2O_2^- + 2H^+ \rightarrow H_2O_2 + O_2$$

Cu-Zn SOD offers protection against damage from reactive oxygen species (ROS) produced as a result of univalent reduction (photoreduction) of dioxygen (O$_2$) in PS I. Through use of rapid freezing and substitution method, followed by innmuno-gold labeling, Ogawa et al. (1995) have shown that in spinach leaf chloroplasts a large proportion of Cu-Zn SOD is attached to the stromal faces of the thylakoids, close to the PS I complex. The proximity of the enzyme to the site of superoxide generation enables rapid disproportionation of the superoxide ions to H_2O_2. Ascorbate peroxidase (APX), which is also bound to the thylakoids in vicinity of the PS I (Miyake et al. 1993), catalyzes the ascorbate based reduction of H_2O_2 to water before it can diffuse into the stroma and cause damage to the enzymes of the CO_2 fixation cycle. It has also been shown that monodehydroascorbate resulting from APX catalyzed oxidation of ascorbate is photoreduced to regenerate ascorbate by the photoreduced ferredoxin in the thylakoids (Miyake and Asada, 1994).

CYTOCHROME C OXIDASE (EC 1.9.3.1)
Cytochrome c oxidase (COX) is a multimetal enzyme protein. Three Cu^+ ions and two haem molecules constitute the metal cofactors of the enzyme. It catalyze the reduction of molecular oxygen (O_2) to H_2O in the terminal step of the mitochondrial electron transport. Cytochrome c oxidase also functions as a proton pump.

$$4 \text{ Cytochrome } c_{(red)} + 4H^+ + O_2 \rightarrow 4 \text{ Cytochrome } c_{(ox)} + 2H_2O$$

OTHER COPPER ENZYMES
The nitrite and nitrous oxide reductases of denitrifying bacteria have copper and haem cofactors.

4.3.2 Copper Proteins

Copper is a constituent of several intensely blue copper proteins such as plastocyanin (Fig. 4.1), which functions as single electron carriers protein.

Plastocyanin
Plastocyanin is a low molecular-mass (11 kDa) blue copper protein universally present in chloroplasts of green plants. More than 50% of the chloroplast copper is bound to this soluble protein. Localized in the lumen of the thylakoid membrane, plastocyanin functions as a mobile electron carrier, linking photosystems II to photosystem I. It accepts an electron from the cytochrome b_6f complex and rapidly donates it to PS I (P700), which functions as a light-dependent plastocyanin-ferredoxin oxidoreductase.

Cu metallothioneins, Phytochelatins
Free ionic copper may bind to low molecular weight proteins in order to form Cu-thioneins (six Cu^{2+} ions mol^{-1}). Copper ions may also bind to polypeptides to form Cu-chelatins. Chelation of copper with thioneins and chelatins provides an effective method for reducing its free (ionic) concentration in the cytoplasm under conditions of copper overload.

4.3.3 Photosynthesis

Copper is involved in the structural organization of PS II (Baron et al. 1990). As a metal constituent of plastocyanin, copper plays a key role in

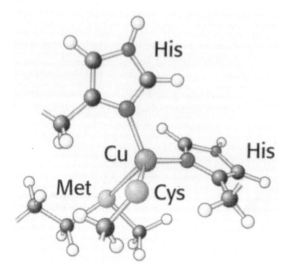

Fig 4.1. Structure of Plastocyanin. The copper atom is liganded to the side chains of the two His residue, a Cys residue and a Met residue in a distorted tetrahedral manner. (Reprinted, with permission, from *Biochemistry.* JM Berg, JL Tymoczko, L Stryer, eds ©2002 W.H. Freeman and company).

linking PS II and PS I and their cooperative functioning in non-cyclic transport of electrons from water to NADP. Plastocyanin accepts electrons from the cytochrome b_6f complex and donates them to PS I, which then transfers them to ferredoxin, enabling it to function as a strong reductant (mid point potential -420 mV), capable of reducing $NADP^+$. Plastocyanin activity is also involved in cyclic transport of electrons associated with PS I mediated ATP production.

Besides its role as a constituent of plastocyanin, copper bound to some polypeptides functions in the maintenance of the lipid environment favouring the movement of plastoquinone molecules during PS II electron transport (Droppa et al. 1984, 1987; Baron et al. 1995; Maksymiec, 1997).

4.4 DEFICIENCY RESPONSES

4.4.1 Visible symptoms

Copper deficiency effects appear first in young, emerging leaves. Symptoms generally initiate as deformation of leaves and cessation of apical growth, followed by development of lateral buds. Symptoms vary with leaf morphology. In graminaceous plants, the youngest leaf fails of unroll and appears needle like. The apical part of the sub-terminal leaves also turn severely chlorotic, withered and twisted. The lower half of these leaves and the old leaves remain apparently normal or turn dark bluish-green. In general, winter cereals like wheat and barley are more sensitive to copper deficiency then the summer crops like maize and sorghum. Copper deficient plants of wheat show profuse tillering, but the tillers remain unproductive and develop symptoms similar to the main shoot. The stem remains slender and upright. Most broad-leaved vegetable crops (cowpea, aubergine) show fine mottling of intervenal areas, rolling of leaf margins and malformations of young leaves. The mottled areas later develop into necrotic patches, particularly along the veins and leaf margins. Leaves of copper-deficient plants often give a wilted appearance, which is attributed to copper involvement in lignification (Rahimi and Bussler, 1973 b; Hopmans, 1990).

In fruit trees, common symptoms of copper deficiency include severe reduction in size and chlorotic mottling of young leaves and marginal scorching of the subterminal (upper middle) leaves. In citrus and deciduous fruit trees, copper deficiency had earlier been described as 'exanthema' or 'die-back'. Spring shoots of Cu-deficient citrus plants appear S shaped and produce abnormally large dark green leaves. Later, these leaves turn chlorotic. Subterminal leaves of the apical shoots develop yellow patches with a diffused outline.

4.4.2 Enzyme Activities

Role of copper as a constituent of enzymes is reflected in a change in activities of enzymes in plants exposed to copper deficiency. Copper-deficient plants show a severe decrease in the activity of Cu-Zn SOD (Ayala and Sandmann, 1988; Tanaka et al. 1995; Yu et al. 1998). There are several reports of decrease in activity of phenolases in leaves of copper-deficient plants (Hewitt and Tatham, 1960; Davies et al. 1978; Judel, 1972; Walker and Loneragan, 1981; Marziah and Lam, 1987). In copper-deficient sunflower, decrease in phenolase activity is associated with increased accumulation of phenolic compounds (Judel, 1972). Walker and Loneragan (1981) showed decreased activity of polyphenol oxidase in subterranean clover leaves to be confined to young leaves, in which copper concentration was reduced to deficiency limits, and which exhibited visible symptoms of copper deficiency. Ascorbate oxidase is another copper enzyme, which is very sensitive to changes in copper concentration. Its activity shows a marked decrease under conditions of copper deficiency (Hewitt and Tatham, 1960; Bar-Akiva et al. 1969; Loneragan et al. 1982; Delhaize et al. 1982; Nautiyal et al. 1999). Nautiyal et al. (1999) reported close correspondence between the level of copper supply, ranging from deficiency to sufficiency, and the activities of ascorbate oxidase and polyphenol oxidase. In view of the close correlation between leaf tissue concentration and ascorbate oxidase activity, ascorbic oxidase has been suggested as a biochemical parameter to diagnose copper deficiency (Bar-Akiva et al. 1969; Delhaize et al. 1982). Activity of copper amine oxidase also shows a close correspondence with the level of copper supply (Delhaize et al. 1986). Bligny and Douce (1977) reported a marked decrease in cytochrome c oxidase activity in mitochondrial preparations of copper-deficient cells of *Acer pseudoplatanus*, with little effect on the respiratory rate. Their findings suggest a possible shift to alternative oxidase pathway in response to copper deficiency stress when cytochrome oxidase may be limiting due to low level of cellular copper.

Copper deficiency also alters the activities of several enzymes which have no specific requirement of copper. Hewitt and Tatham (1960) reported a marked increase in the activities of peroxidase, acid phosphatase and isocitrate dehydrogenase in leaves of copper-deficient tomato plants. Activities of peroxidase and IAA oxidase are also reported to be enhanced under copper-deficiency (Davies et al. 1978).

4.4.3 Photosynthesis

Copper deficiency results in decreased photosynthetic activity (Spencer and Possingham, 1960; Botrill et al. 1970; Barr and Crane, 1974; Horvath et al. 1983; Casimiro et al. 1990; Ayala et al. 1992; Pandey and Sharma, 1996). The most obvious cause of this is a decrease in photosynthetic electron transport activity (Baszynski et al. 1978; Droppa et al. 1984; Casimiro et al. 1990). While activities of both PS II and PS I are inhibited, the effect is more pronounced on PS I. Fluorescence measurements on intact flag leaves of wheat plants show more severe inhibition of PS I activity than that of PS II, with associated decrease in PS I / PS II activity ratio. This reconciles with the role of copper as a constituent of plastocyanin, which couples the transfer of electrons from the cytochrome complex of PS II to the PS I reaction centre (Katoh, 1977; Sandmann et al. 1983). In some algae and cyanobacteria. synthesis of plastocyanin is dependent on the availability of copper (Merchant and Dreyfuss, 1998). Under copper starvation, *Chlamydomonas reinhardtii* synthesizes a haem protein-cytochrome c_6, which substitutes for plastocyanin to function as carrier of electrons from the cytochrome complex of PS II to PS I reaction centres (Merchant and Dreyfuss, 1998). Shikanai et al. (2003) suggested that under conditions of copper deficiency, plastocyanin-mediated electron transport and chloroplastic Cu-Zn SOD activity may suffer a loss because of limited availability of copper within the chloroplast. Based on characterization of six *Arabidopsis* mutants defective in *PAA 1* gene, Shikanai et al. (2003) have concluded that the *Arabidopsis* gene *PAA 1* encodes a P-type ATPase (*PAA 1*), which mediates the transport of copper across the plastid envelope. Inspite of adequate leaf tissue copper, the chloroplastic concentration of copper in the *paa 1-1* mutant may not be enough to meet the cofactor requirements of plastocyanin and Cu-Zn SOD. The electron transport defect of *paa 1-1*, evident under copper deficiency ($< 1\mu M$ Cu), could be suppressed on elevating the level of copper supply ($10 \mu M$ Cu).

Copper deficiency may also contribute to decreased photosynthetic activity because of its effect on chloroplast organization, photosynthetic pigments and leaf CO_2 conductance. Photosynthetic electron transport may be inhibited due to copper deficiency effect on the plastoquinone content of the thylakoids (Droppa et al. 1987; Horvath et al. 1987), possibly because of copper involvement in polyphenol oxidase, which contributes to synthesis of quinones. Droppa et al. (1984) reported that copper is involved in the synthesis of polypeptides that contribute to the fluidity of the chloroplast membranes and provide a proper lipid environment for plastoquinone-mediated electron transfer. Droppa et al. (1989) and Baron et al. (1993) have attributed low rates of PS II electron transport in Cu-deficient plants to Cu-deficiency induced changes in lipid composition of the thylakoid membranes. The thylakoid lipids of the Cu-deficient plants contain high levels of unsaturated fatty acids, causing impairment in PS II electron transport.

Copper-deficient plants showed ultrastructural changes in the chloroplasts (Baszynski et al. 1978; Henriques, 1989; Casimiro, 1990). Transmission electron microscope examination of flag leaf chloroplast ultrastructure in copper-deficient wheat plants (Casimiro et al. 1990) show disorganization of the thylakoids at the level of grana. While chloroplasts of copper-sufficient plants show well-structured grana and intergrana, the copper-deficient chloroplasts exhibit disorganized and swollen thylakoids. Copper-deficient plants also show a decrease in the concentration of photosynthetic pigments (Baszynski et al. 1978; Horvath et al.1983; Henriques, 1989; Pandey and Sharma, 2002a). Copper deficiency induced decrease in chlorophyll content (Casimiro et al. 1990; Pandey and Sharma, 2002a) may not be limiting for photosynthesis during early stages of deficiency, but prolonged and acute deficiency of copper causes substantial loss of photochemical activity of chloroplasts (Pandey and Sharma, 1996).

Decrease in leaf conductance of CO_2 in copper-deficient plants (Casimiro, 1987; Pandey and Sharma, 1996, 2002a) may lower the CO_2 concentration in mesophyll cells to levels that may further affect photosynthetic efficiency. Pandey and Sharma (1996) showed that restoration of adequate supply of copper to safflower (*Carthamus tinctorius*) plants reversed the decrease in both leaf conductance and rate of photosynthesis. (Table 4.1)

4.4.4 Carbohydrates

Copper deficiency affects the leaf tissue concentration of soluble carbohydrates but the increase or decrease changes with the stage of plant development (Graham, 1980). During the vegetative phase, copper-deficient

Table 4.1. Copper effect on leaf conductance, photosynthesis and transpiration in safflower (*Carthamus tinctorius*) leaves

	Plant copper status		
	Cu-sufficient	Cu-deficient	Cu-resupplied
Tissue copper (μg g^{-1} dry wt.)	8.0	3.4	5.1
Stomatal conductance (m mol m^{-2}s^{-1})	3.83	1.54	2.73
Intercellular CO_2 (μg l^{-1})	3.50	3.19	4.24
Hill activity (DC PIP reduced mg^{-1} chl.)	1.24	0.49	0.61
Photosynthetic rate (μ mol CO_2 m^2 s^{-1})	5.89	3.91	4.08
Transpiration rate (μg^{-1} c^{-2} s^{-1})	6.99	2.48	4.40

(After Pandey and Sharma, 1996)

plants show decreased concentration of soluble sugars (Brown and Clark, 1977; Mizuno et al. 1982), which is in consonance with low rates of photosynthesis under copper deficiency (Casimiro et al. 1990; Ayala et al. 1992). Copper deficiency response during the reproductive phase may be just the opposite, at least in cereal crops, where developing grains form the dominant sink for carbohydrates and the sink capacity is severely limited by copper deficiency (Section 4.4.7). In wheat plants, wherein setting of seeds is severely limited by copper deficiency, flag leaf shows high accumulation of soluble carbohydrates (Graham, 1980). Development of many non-productive tillers during the reproductive phase of the copper-deficient plants has been attributed to utilization of the accumulated carbohydrates for renewed vegetative growth.

4.4.5 Phenolics, Lignification

Leaves of copper-deficient plants show increased accumulation of phenolic substances (Judel, 1972; Adams et al. 1975; Robson et al. 1981). Judel (1972) reported that increased accumulation of phenolic compounds in leaves of copper-deficient sunflower plants is associated with a concomitant decrease in the activity of polyphenol oxidase, which catalyzes the oxidation of phenolic compounds. Robson et al. (1981) made a study of the phenolic constituents of young leaves of wheat plants subjected to and recovering from copper deficiency. Copper deficiency caused an increase in the concentration of phenolic acids including ferulic acid, but decreased the concentration of p-coumaric acid. The former was attributed to their limited polymerization to form lignins, possibly because of decrease in laccase activity. In wheat, copper deficiency reduces lignification and thickening of cell walls of xylem elements, fibres and epidermal cells (Schutte and Mathews, 1969). Histochemical localization of lignin (using acidic phloroglucinol) revealed poor lignification of xylem vessels under copper deficiency (Rahimi and Bussler, 1973b). The wilted appearance of young leaves of well-watered but copper-deficient plants is attributed to poor lignification of the cell walls (Rahimi and Bussler, 1973a) and so is the case with the failure of rupture of the anthers of copper-deficient plants along the stamina (Dell, 1981).

4.4.6. Stress Disposition

As a constituent of Cu-Zn superoxide dismutase, copper functions in the disproportionation of superoxide (O_2^-) ions. Localized close to the PS I complex (Ogawa et al. 1995), Cu-Zn SOD catalyzes rapid dismutation of the superoxide ions to generate hydrogen peroxide (H_2O_2) and molecular oxygen (O_2). Decrease in Cu-Zn SOD activity resulting from inadequate supply of copper weakens the antioxidant defence system of plants, making them susceptible to oxidative damage.

Copper involvement in tissue lignification (Graham and Webb, 1991) contributes to a protective role of copper against pathogenic infections (Graham, 1983). Decrease in lignin content in epidermal cells reduces the resistance to penetration of the pathogenic organisms and makes copper deficient plants more vulnerable to pathogenic infections. Copper deficiency makes wheat plants susceptible to take-all. Soil application of copper suppresses the infection. Protection against pathogens may also be provided through enhanced generation of reactive oxygen species, which are cytotoxic to the pathogen and also activates the defense mechanism of the host (Lamb and Dixon, 1997). Rea et al. (1998, 2002) have shown a role of the copper diamine oxidases in activation of defense mechanism of plants. In chickpea, wounding and pathogenic infection (*Ascochyta rabiei*) activate copper amine oxidase leading to enhanced production of H_2O_2, which contributes to the defense mechanism of the host (Lamb and Dixon, 1997).

4.4.7 Reproductive Development

Copper is involved in plant reproductive development in more than one way. Copper deficiency leads to delayed flowering and a reduction in the number of flowers (Graves and Sutcliffe, 1974; Davies et al. 1978; Reuter et al. 1981; Pandey and Sharma, 2002a). The latter can often be attributed to decrease in the number of flowering shoots. In cotton, where flowers are borne on the tertiary branches, flowering is drastically reduced because of failure of emergence of these branches in plants deprived of adequate copper supply (Pandey and Sharma, 2002a). Resupply of copper to copper-deficient plants during vegetative phase initiates renewed growth including development of tertiary branches, which contributes to an increase in the number of flowers. One of the reasons for delayed flowering of copper-deficient plants (Davies et al. 1978; Reuter et al. 1981) has been attributed to a decrease in the activity of polyphenol oxidase, leading to increased accumulation of phenolics, and IAA oxidase and peroxidases, causing the accumulation of IAA to levels that may become inhibitory to flowering.

Higher concentration of copper in anthers and ovaries than in other floral parts (Knight et al. 1973) suggests copper involvement in development of reproductive parts. Inadequate supply of copper limits the size of anthers (Graham, 1975; Agarwala et al. 1980; Dell, 1981) and prevents their dehiscence (Dell, 1981). Failure of anther dehiscence in copper-deficient plants is attributed to poor lignification of anther cell walls, which is critical for the rupture of stamina and, thus, release of pollen grains.

The anthers of copper-deficient plants show a non-functional tapetum (Jewell, 1988). Instead of supplying nutrients to the developing pollen grains, the tapetum becomes expansionary. Copper-deficient plants of durum wheat are reported to show abnormal (polyploid) pollen grains (Azouaou and Souvré, 1993). In copper-deficient wheat plants, the number and size of pollen grains is severely restricted, pollen grains lack dense cytoplasmic contents, appear deflated and lack fertility (Graham, 1975; Agarwala et al. 1980). The pollen grains of copper-deficient plants lack the high starch content that is characteristic of normal, viable pollen grains (Agarwala et al. 1980; Jewell et al. 1988). *In vitro* germination of copper-deficient pollen grains is severely limited (Agarwala et al. 1980). Pandey et al. (1995a) followed ultrastructural changes in pollen grains of green gram following copper deprivation and resupply. Copper deficiency caused a decrease in the size and pore diameter of pollen grains and increase in the thickness of exine. The exine architecture was also altered. The muri were thick and with undulated walls and the baculae showed pilate projections. In the pollen grains produced following resupply of Cu, the ultrastructural changes were minimized and germination percentage was increased. Cross-pollination studies by Graham (1975) showed that decreased seed setting in copper-deficient wheat plants is a function of male sterility (and not ovule sterility). Pollination of copper-sufficient plants with pollen grains of copper-deficient plants limited the number of grains head[-1] to just 2, as against 47 in copper-deficient plants fertilized with pollen grains of copper-sufficient plants (Graham, 1975). Role of copper in microsporogenesis and pollen fertility offers an explanation for more severe reduction in reproductive (seed) yield than dry matter production by copper-deficient plants (Brown et al. 1972; Graham, 1975). Copper also affects the maturation and quality of seeds (Nautiyal and Chatterjee, 1999; Nautiyal et al. 1999). Copper-deficient rice plants are reported to show increased activities of invertase and amylase during the grain filling stage and decrease in seed reserves of starch, soluble sugars and protein (Nautiyal et al. 1999).

4.4.8 Water Relations

Copper status of plants affects water relations parameters (Graham, 1976; Sharma and Sharma, 1987; Pandey and Sharma, 1996, 2002a). Leaves of copper-deficient plants of safflower have low stomatal conductance associated with low rates of transpiration and these changes are largely reversed on elevating copper supply from deficiency to sufficiency (Fig. 4.1).

CHAPTER 5

ZINC

5.1 GENERAL

Zinc is essential for all organisms. It is a group II b metal, with a completed d subshell and two additional s electrons. Thus, unlike other cationic micronutrients, it has only one oxidation state (Zn^{2+}). Zinc plays a structural as well as functional role in plants. It forms a structural component of a large number of proteins with catalytic or regulatory functions. Usually, the Zn^{2+} ion binds to nitrogen or sulphur-containing ligands through ionic bonds, forming a tetrahedral geometry. Over three hundred enzymes are known to contain Zn as a cofactor (Valee and Auld, 1990). By providing stability to many regulatory proteins, or domains thereof (zinc fingers, zinc clusters and RING finger domains), zinc plays a role in transcriptional regulation (Coleman, 1992). Zinc may chelate with polypeptides, synthesized in response to excessive accumulation of heavy metals (including itself), to form metallopolypeptides like phytochelatins and metallothioneins and contribute to tolerance mechanisms against metal hyperaccumulation. Over the years, zinc has been shown to play critical roles in plant reproductive development, prevention of water stress and protection against toxic effects of reactive oxygen species.

5.2. UPTAKE, TRANSPORT, DISTRIBUTION

Zinc is largely taken up by plants in the ionic form, as free Zn^{2+}. Studies based on root uptake of Zn^{2+} by certain graminaceous species grown over a wide range of Zn concentrations in solution cultures (Chaudhry and Loneragan, 1978; Veltrup, 1978; Mullins and Sommers, 1986) showed Zn^{2+} uptake to be concentration-dependent and saturable, which implied its carrier-mediated transport. Measurement of Zn^{2+} influx from chelated buffered solutions, in which Zn^{2+} activity was maintained close to the concentrations found in soil solution in most agricultural soils, lent support to carrier-mediated transport of Zn^{2+} (Norvell and Welch, 1993; Hart et al. 1998). Hart et al. (1998) showed Zn^{2+} uptake by wheat seedlings to be concentration dependent, with smooth saturating curves characteristics of carrier-mediated transport. Uptake was also shown to be inhibited by low temperatures indicating dependence on active metabolism. Zinc uptake measurements, using the *Chara corellina* model (Reid et al. 1996) showed a bi-phasic pattern of uptake, which showed that Zn^{2+} uptake was mediated by two separate systems—a high-affinity system that saturated at 100 nm and a low-affinity system that showed a linear dependence on concentrations upto at least 50 mM. Strong evidence in support of involvement of both a low concentration–high affinity system and a high concentration–low affinity system in Zn^{2+} uptake has come from short-term measurement of $^{65}Zn^{2+}$ uptake by wheat cultivars grown hydroponically at zinc concentrations ranging from 0.1 nM to 80 mM (Hakisalhoglu et al. 2001). According to Hakisalihoglu et al. (2001), the high-affinity transport system forms the predominant mechanism for Zn^{2+} uptake from soils, in which Zn^{2+} concentration in the soil solution lies close to the lower limits of the nanomolor range (Welch, 1995).

Phytosiderophores, involved in iron uptake by graminaceous plants, are also secreted by the plants in response to Zn-deficiency (Zhang et al. 1989, 1991a; Walter et al. 1994; Cakmak et al. 1994, 1996; Rengel 1997; Rengel et al. 1998; Hopkins, 1998). Reported ability of phytosidophores to mobilize not only iron but also zinc from calcareous soils (Treebly et al. 1989) suggested their possible involvement in mobilization and uptake of zinc under Zn-deficiency conditions. Evidence for a role of phytosiderophores in enhancing Zn mobilization and uptake came for the studies reported by Zhang (1999a) and Von Wiren et al. (1996) and reports of a positive relationship between the ability of wheat genotypes to secrete phytosiderophores and their zinc efficiency (Walter et al. 1994; Cakmak et al. 1994, 1996). Hopkins et al. (1998) attributed the differences in tolerance of sorghum, wheat and corn to Zn deficiency to the ability of their roots to release phytosiderophores under conditions of zinc deficiency. When Zn supply was limiting, release of phytosiderophores by roots followed the order wheat> sorghum >corn, which corresponded to their tolerance to Zn deficiency (Hopkins et al. 1998). Doubts have, however, been raised about release of phytosiderophores by roots of graminaceous plants as a mechanism for enhanced mobilization and uptake of Zn in response to Zn deficiency. Erenoglu et al. (1996) did not find any positive relationship between Zn efficiency of bread wheat genotypes and the capacity of their roots to secrete phytosiderophores. According to Rengel (1998), enhanced secretion of phytosiderophores under Zn deficiency is possibly caused by Zn-deficiency induced impairment in iron transport resulting in its deficiency, which activates the secretion of the phytosiderophores.

Several transporters belonging to the *ZIP* (*ZRT-IRT*-like proteins) (Guerinot, 2000) and the CDF (cation diffusion facilitator) family (Williams et al. 2000) have been recently suggested to be involved in transport of zinc. Four ZIP transporter genes, *ZIP1, 2, 3* and *4*, have been isolated from *Arabidopsis* (Grotz et al. 1998; Guerinot, 2000). *ZIP1, 2, 3* and *4* are expressed in roots of zinc-deficient plants. *ZIP4* is also expressed in shoots and is predicted to have a chloroplast-targeting sequence. Expression of *ZIP1, ZIP2,* and *ZIP3* genes of *Arabidopsis* in zinc-defective yeast mutant *zrt1 zrt2* has been shown to restore growth limitation due to zinc-deficiency. Another zinc transporter gene ZNT1 has been cloned from Zn/Cd accumulator plant *Thalspi caerulescens* (Lasat et al. 2000; Assuncao et al. 2001). The Tc ZNT1 is a ZIP gene homologue and expressed at high levels in both roots and shoots (Assuncão et al. 2001). In the non-hyperaccumulator species *T.arvense, TcZNT1,* and another ZIP gene homologoue ZNT2, are expressed at much lower levels and only in response to zinc deficiency (Lasat et al. 2000; Assuncão et al. 2001). Another member of the ZIP family *Gm ZIP1,* isolated from soybean, has recently been shown to functionally complement the *zrt1/zrt2* mutant of yeast (Moreau et al. 2002). It has been reported that *Gm ZIP1* is highly selective for zinc and is expressed only in root nodules, where it is largely localized to the peribacterial membrane, suggesting its possible role in symbiosis (Moreau et al. 2002).

Van der Zaal et al. (1999) have characterized a cation diffusion facilitator family transporter ZAT from *Arabidopsis* and showed that it functions in zinc transport under condition of zinc excess. *At ZAT* is constitutively expressed throughout the plant. Its overexpression in transgenics exposed to zinc excess leads to increased accumulation of zinc. Van der Zaal et al. (1999) have suggested that ZAT-mediated transport of zinc leads to vascular/vacuolar sequestration of zinc, contributing to zinc homeostasis and tolerance.

Zinc is phloem mobile. Haslett et al. (2001) have shown that foliar application of zinc to wheat can provide sufficient zinc for meeting the requirements for vigourous vegetative growth. Zinc, supplied through leaves, was shown to be phloem translocated to different plant parts including roots. In wheat, sufficient amount of zinc (^{65}Zn) has been shown to be translocated from leaves to the developing grains during the grain-filling period (Pearson and Rengel, 1994, 1996; Pearson et al. 1995). Transport to seeds has been shown to be influenced by zinc status of the mother plants (Pearson and Rengel, 1995a). Limitation in zinc transport under conditions of zinc deficiency could result from its possible binding to cellular metabolites, Zn enzymes and the exterior of the plasma membrane (Welch, 1995).

Phloem transport of Zn takes place in the form of zinc complexes with organic acids such as citric and malic acids (White, 1981a,b) or nicotianamine (Stephan and Scholz, 1993; Schmidke and Stephan, 1995) Recent studies by Takahashi et al. (2003) suggest a role of nicotianamine in transport of zinc to floral parts and intracellular trafficking to sites where it is integrated in to the functional proteins.

5.3 ROLE IN PLANTS

5.3.1. Enzyme Action

Zinc is a constituent of a multitude of enzymes (Valee and Auld, 1990; Valee and Falchuk, 1993; Berg and Shi, 1996). Since zinc has only one oxidation state (Zn^{2+})—unlike other metalloenzymes—Zn enzymes cannot function in oxidation-reduction reactions. In the Zn metalloproteins, the zinc ion binds to three-imidazole groups (Schiffs' base), with the fourth coordination position left free for interacting with the substrate. Generally, the fourth coordination site is bound to a water molecule that facilitates hydration or hydrolysis reactions and most zinc enzymes catalyse such reactions.

CARBONIC ANHYDRASE (EC 4.2.1.1)

Carbonic anhydrase (CA) contains zinc as an essential cofactor and is found in abundance in all living beings. Unlike the mammalian enzyme, which is a monomer, the plant enzyme can be a dimer, tetramer, hexamer or an octamer. Each subunit contains a single zinc atom, bound close to the active site to three imidazole rings of the histidine residues. The fourth coordination site is left free to react with the substrate. The enzyme catalyzes the reversible conversion of carbon dioxide to bicarbonate, which has high solubility in water.

$$CO_2 + H_2O \rightleftharpoons HCO_3^- + H^+$$

Carbonic anhydrase is critical to photosynthesis in C_4 plants, in which the first carboxylation reaction is catalyzed by phospho*enol*pyruvate carboxylase (PEPCase), which uses bicarbonate as a substrate (O'Leary, 1982). Unlike the C_3 plants, wherein CA is largely localized to mesophyll chloroplasts, in C_4 plants CA is localized to the cytosol of the mesophyll cells, which is also the site for PEPCase-catalyzed carboxylation (Burnell and Hatch, 1988). An analysis of the CA activity (at *in vitro* CO_2 concentrations) and associated rates of photosynthesis, showed that even a small decrease in CA activity could result in such a large decrease in steady-state HCO_3^- concentration, as could limit C_4 photosynthesis (Hatch and Burnell, 1990). In C_3 plants, the role of CA in photosynthesis has been somewhat uncertain (Utsunomiya and Muto, 1993). Randall and Bouma (1973) observed decreased CA activity with little effect on photosynthesis in zinc-deficient plants. Inhibition of CA activity did not cause a decrease in photosynthesis (Sasaki et al. 1998).

SUPEROXIDE DISMUTASE (EC 1.1.1.2)
The enzyme structure and catalytic function have been described in Chapter 4 (Section 4.3.1.).

ALCOHOL DEHYDROGENASE (EC 1.1.1.2)
Alcohol dehydrogenase contains two zinc atoms per molecule, one of which performs a structural and the other a catalytic function (Coleman, 1992) Alcohol dehydrogenase catalyzes the oxidation of acetaldehyde, formed by decarboxylation of PEP generated during glycolysis, to ethanol.

$$\text{Acetaldehyde} + \text{NADH} \rightleftharpoons \text{Ethanol} + \text{NAD}^+$$

Increased activity of alcohol dehydrogenase in response to anaerobic stress such as flooding enables the plant roots/tissues to temporarily meet the energy requirement from ethanolic fermentation (Gibbs and Greenway, 2003; Ravichandran and Pathmanabhan, 2004).

CARBOXYPEPTIDASE A (EC 3.4.17.1)
Carboxypeptidase A is a zinc metalloenzyme that catalyzes the hydrolysis of peptide bond at the C-terminal end by activating a water molecule.

$$\text{Peptidyl-L-amino acid} + H_2O \rightleftharpoons \text{Peptide} + \text{amino acid}$$

Carboxypeptide A is widely distributed. The plant enzyme catalyzes the hydrolysis of reserve proteins of seeds.

Other Zinc Enzymes

Zinc is a cofactor of certain DNA-dependent RNA polymerases involved in transcription (Falchuk et al. 1977; Petranyl et al. 1978). Zinc is tightly bound to the enzyme protein and its removal results in enzyme inactivation. Petranyl et al. (1978) reported inactivation of wheat germ RNA polymerase II on removal of Zn from the enzyme protein. Strater et al. (1995) have described the crystal structure of a purple acid phosphatase from kidney bean which contains a dinuclear Fe (III)-Zn(II) active site.

There are several enzymes that are known to have a cofactor requirement of zinc in animals, fungi and bacteria but not in plants. Some of these are of wide occurrence in plants and their activities are inhibited under zinc deficiency. Common examples of such enzymes and the reactions catalyzed by them are:

NAD+-Glutamate Dehydrogenase (EC 1.4.1.2)
$$\text{L-Glutamate} + \text{NAD}^+ \rightleftharpoons \alpha\text{-Ketoglutarate} + NH_4^+ + \text{NADH} + H^+$$

Aldolase (EC 4.1.2.13)
$$\text{Fructose-1,6-bisphosphate} \rightleftharpoons \text{Dihydroxy acetone phosphate} + \text{D-Glyceraldehyde-3-phosphate}$$

ALA Dehydratase (EC4.2.1.24)
$$\delta\text{-Aminolevulinate} \rightleftharpoons \text{Porphobilinogen}$$

H⁺-ATPase

A plasma membrane H^+-ATPase of corn roots has been shown to use zinc as the substrate and function as Zn-ATPase (Katrup et al. 1996).

5.3.2 Zinc Finger Proteins

Apart from being a constituent of zinc-metalloenzymes, zinc functions as a constituent of several regulatory proteins which interact with DNA and control gene expression. In the form of zinc finger, it forms a structural motif of the DNA-binding region of the transcriptional regulatory proteins (Berg and Shi, 1996). Originally recognized in transcription factor TF IIIA, involved in the transcription of 5s RNA by RNA Polymerase III genes in *Xenopus*, zinc finger motifs have been identified for several other transcription factors (Klug and Rhodes, 1987). The TF IIIA-type DNA binding proteins, also designated as the C2H2 or the classical zinc finger proteins, contain a Zn atom which is tetrahedrally coordinated to two Cys and two His residues to form a compact finger structure (Fig. 5.1). Structural studies over the past decade have shown that the TF IIIA-type Zn finger proteins of plants have two characteristic features, which distinguish them from other eukaryotic zinc figure proteins. First, the α - helix of the zinc finger, which makes contact with the major genome of DNA, contains a conserved amino acid sequence QALGGH; second, the adjacent fingers are separated by long spacers of diverse lengths, which recognize the spacing between the core sites of the target DNA (Takatsuji, 1999). The number of zinc fingers in a TF IIIA-type zinc finger protein range from one to four, with each additional finger resulting in increased specificity (Takatsuji, 1999; Laity et al. 2001). Some TF IIIA-type zinc finger proteins are involved both in protein-protein and protein-DNA interactions (Mackay and Crossby, 1988). While some zinc fingers interact with proteins, others react with DNA. First to be identified in a DNA-binding protein of petunia (EPF, renamed ZPT2-1), (Takatsuji et al. 1992), several TF IIIA-type zinc finger proteins have been isolated and characterized from different plants species and shown to be implicated in floral organogenesis, leaf initiation, gametogenesis and stress responses (Takatsuji, 1999). The gene encoding a TF IIIA-type zinc finger protein *SUPERMAN* (*SUP*), isolated and characterized from *Arabidopsis* has been shown to control the cell division demarcating the boundary between the third and the fourth whorl of flowers and determine the stamen number and carpel development (Sakai et al. 1995). It is also involved in morphogenesis (Gaiser et al. 1995). Nakagwara et al. (2004) have described a *SUP* gene homologue from petunia (*Ph SVP 1*), whose product also functions in morphogenesis of placenta, anther and connective tissues between the floral organs. More recently, Nakagwara et al. (2005) have described another zinc finger protein from petunia which structurally resembles SUP and designated it as LIF (for Lateral shoot-Inducing Factor).

The LIF cDNA has been shown to be specifically expressed around the dormant axillary buds of transgenic petunia plants. Its over expressions induces axillary buds and dwarfed growth. The alterations in plant forms are attributed to changes in the cytokinin metabolism. Nakagawa et al. (2005) haves suggested that controlled expression of LIF transgene can possibly be made use of in optimizing the extent and the pattern of shoot branching to increase the harvest yield of plants.

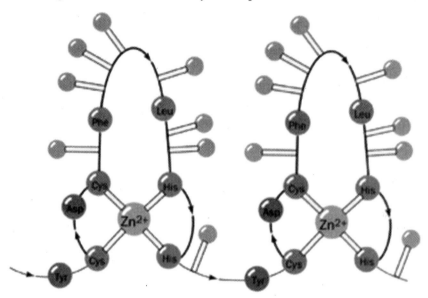

Fig 5.1. Schematic diagram of zinc finger motifs. Each unit shows a Zn^{2+} ion tetrahedrally liganded to two Cys and two His residues. (Reprinted, with permission, from *Biochemistry*. D Voet, JG Voet, eds ©2004 John Wiley & Sons. Inc).

Yanagisawa (1995) have characterized a family of single zinc finger proteins, named Dof (for DNA binding with one finger), from plants. In the Dof zinc finger protein, the zinc atom is coordinated to Cys residues only (C2C2), with the zinc finger domain containing a 52 amino acid stretch. The Dofs are suggested to function as transcriptional activators or repressors of tissue-specific and light-regulated gene expressions in plants (Yanagisawa and Sheen, 1998).

Recently, transcription factors with zinc-finger binding domains have been designed and used for targeting specific DNA sequences of endogenous genes (Guan et al. 2002; Ordiz et al. 2002). Guan et al. (2002) designed a polydactyl zinc finger transcriptional factor that could target the *Arabidiopsis* gene *APETALA 3 (AP 3)* and induce a dramatic change in its expression. Another important biotechnological use to which zinc finger proteins have been put is the designing of zinc finger nucleases and using them for targeted mutagenesis (Lloyd et al. 2005). Zinc finger nucleases, synthesized by fusing non-specific DNA cleavage domains from *Fok* 1 endonuclease with DNA–binding domains composed of three Cys-2 His-2 fingers (Kim et al. 1996), make effective tools for targetted mutagenesis of plant genes (Lloyd et al. 2005).

5.3.3 Membrane Integrity

Bettger and O'Dell (1981) showed the importance of zinc in maintaining the structural integrity of biomembranes. Welch et al. (1982) were the first to suggest zinc involvement in the permeability of plant plasma membranes. They showed enhanced leakage of ^{32}P and ^{36}Cl from roots of wheat plants subjected to zinc deficiency. Resupply of zinc showed reversal of enhancement in leakage of the ions. Welch et al. (1982) concluded that deficiency of zinc makes the plasma membranes more leaky due to loss of structural integrity. Lindsay et al. (1989) suggested that zinc is involved in protection of the plasma membrane at its apoplasmic side. They opined that zinc reacts with negatively charged molecules of the membranes and contributes to their structural stability. Cakmak and Marschner (1988a) showed enhanced leakage of solutes (K^+, amino acids, sugars, phenolics) by roots of zinc-deficient plants of four different plant species, which could be reversed on resupply of zinc, a feature not shown by calcium that is known for its effect on membrane permeability. Plasma membrane vesicles isolated from roots of zinc-deficient plants also showed higher permeability (proton flux) than the vesicles isolated from roots of Zn-sufficient plants (Pinton et al. 1993).

Effect of zinc deficiency in increasing membrane permeability resembles the effect induced by enhanced peroxidation of the membrane constituents (Van Ginkel and Sevanian, 1994). Zinc-deficient plants show decreased fatty acid content of the membranes. Effect on the concentration of unsaturated fatty acids and phospholipids (Cakmak and Marschner, 1988c) and that on reactive sulphydryl (-SH) groups is particularly marked (Welch and Norvell, 1993; Rengel, 1995b). The decrease in the concentration of -SH groups under zinc deficiency is more in zinc-sensitive wheat cultivar Durati, than in tolerant cultivar Warigal, and is reversed on resupply of zinc (Rengel, 1995b). Zinc has been suggested to protect the -SH groups of plasma membrane proteins from oxidative damage.

5.3.4 Anti Oxidant Activity

There is increasing evidence for a protective role of zinc against oxidative stress (reviewed by Cakmak, 2000). Oxidative stress results from production of reactive oxygen species (ROS) in quantities in excess of what can be effectively detoxified by the plant's inherent antioxidant system. Zinc protects plants from oxidative damage in two ways: by preventing enhanced production of the ROS; and by their rapid detoxification. Zinc inhibits the accelerated production of ROS by inhibiting the activity of membrane bound NADPH oxidase, which catalyzes the production of superoxide ions.

$$2O_2 \longrightarrow 2O_2^-$$
$$NADPH\ oxidase$$

Bettger and O'Dell (1981) provided evidence of zinc involvement in protection of cellular membranes against peroxidative damage from NADPH-derived free radicals. Enhanced peroxidation of membrane lipids destabilizes the membranes and makes them more leaky.

As a constituent of superoxide dismutase (Cu, Zn SOD), a key component of the inherent antioxidant defense mechanism of plants, zinc catalyzes rapid detoxification of superoxide ions by producing hydrogen peroxide, which can be taken care of by the other enzymatic components of the antioxidant system (e.g. catalase, ascorbate peroxidase). Accumulation of superoxide ions (O_2^-) may lead to production of hydroxyl ($OH^{\bullet}$) radicles (Haber-Weiss reaction), which are even more reactive than O_2^- and cause more severe damage to cellular membranes, nucleic acids and proteins.

5.4 DEFICIENCY RESPONSES

5.4.1 Visible Symptoms

Zinc deficiency symptoms generally appear in subterminal leaves after apparently normal growth during the early stages. Zinc deficiency responses of plants are varied. Common symptoms, shared by most plants, include reduction in leaf size and condensation of shoot growth. Combination of the two is often reflected as the clustered or rosetted appearance of the terminal growths. Fading of lamina, intervenal chlorosis, often associated with development of reddish-brown or bronze tints, and a metallic sheen on leaves are other common symptoms of zinc-deficiency. In some crops, leaves of zinc-deficient plants exude a viscous fluid, which on drying, appears as salt incrustations on the leaf surfaces. Such a feature is possibly a result of damage caused to cell membranes (Section 5.3.3). Response of vegetable crops to zinc deficiency is varied. Beans and radish exhibit zinc deficiency in the form of intervenal chlorosis. The chlorotic leaf areas turn brown, papery and necrotic, ending up in disintegration of the lamina. Leaves of zinc-deficient plants of cabbage and cauliflower show thickening of laminae and a thick deposition of epicuticular wax. These leaves remain turgid, possibly because of restricted opening of stomata. Leaves of zinc-deficient tomato plants show epinasty and inward curling of lamina, resembling auxin deficiency response. The laminae also turn thick and brittle. Older leaves often develop brown or orange tints and turn necrotic.

Amongst the cereal food crops, maize, rice and wheat are extremely sensitive to zinc deficiency. Symptoms initiate in fully expanded young leaves as fading of lamina and appearance of light brown necrotic lesions some distance away from the base. With continued deficiency, the lesions coalesce and spread apically. The apical part of the leaves of Zn-deficient wheat plants wither out, curl, hang down and even break out from the middle of the leaf. Zn-deficient wheat leaves also show epidermal exudations suggestive of leaky membranes.

5.4.2 Enzyme Activities

Plants subjected to zinc deficiency display alterations in the activity of many enzymes, including those which are not known to have any specific requirement of zinc as a cofactor or activator. Decrease in carbonic anhydrase activity in leaves is an universal response to zinc deficiency (Wood and Sibly 1952; Bar-Akiva and Lavon, 1969; Edward and Mohamad, 1973; Sharma et al. 1981a, 1982; Seethambaram and Das 1985; Silverman 1991; Pandey and Sharma, 1998, 2000; Rengel, 1995; Sasaki et al. 1998; Pandey et al. 2002b). Resupply of zinc to Zn-starved plants reverts the enzyme activity to normal (Sharma et al. 1981a; Bisht et al. 2002b). While there is a correlation between zinc supply, leaf tissue concentration of zinc and carbonic anhydrase (CA) activity, for the same level of zinc content different genotypes may show substantial differences in CA activity. Rengel (1995a) described higher CA activity in Zn-efficient wheat genotype 'Warigal' than in Zn-inefficient wheat genotype 'Durati' even when their leaf tissue concentration of zinc was about the same. Zinc deficiency induced decrease in CA activity has been shown to be associated with a decrease in the stomatal opening (Sharma et al. 1995; Pandey and Sharma, 2000) and an increase in CO_2 transfer resistance (Pandey and Sharma, 1989; Sasaki et al. 1998). Sharma et al. (1995) showed a good correlation between zinc concentration, guard cell CA activity (and K^+ content) and stomatal opening. Zinc-deficient rice plants are reported to show loss of alcohol dehydrogenase activity (Moore and Patrick, 1988).

Zinc-deficient plants also show decreased activity of superoxide dismutase (SOD) (Cakmak and Marschner, 1993; Yu et al. 1998; Yu and Rengel, 1999; Pandey et al. 2002a,b). As expected, compared to the total SOD activity, the activity of Cu-Zn SOD is more responsive to zinc deficiency (Yu and Rengel, 1999; Pandey et al. 2002a,b). Leaves of zinc-deficient plants, showing low Cu-Zn SOD activity, may exhibit elevated activity of Fe-SOD (Yu and Rengel, 1999 Bonnet et al. 2000). As in case of CA, genotypes may differ in their efficiency of zinc utilization in SOD activity. Cakmak et al. (1997c) showed genotypic differences in SOD activity for the same level of zinc deficiency. Activities of other enzymes of the antioxidative defense system are also influenced by Zn nutrient status (Fig. 5.2). Zinc-deficient plants show low activities of catalase (Marschner and Cakmak, 1989; Bisht et al. 2002b) and ascorbate peroxidase (Yu et al. 1998; Pandey et al. 2002b).

Glutathione reductase (GR), an important enzyme of the ascorbate-glutathione cycle, also responds to changes in Zn nutrient status of plants but the effects of Zn deficiency on GR activity are inconsistent. Cakmak and Marschner (1993) described a decrease in GR activity in zinc-deficient bean plants. On the contrary, Pandey et al. (2002b) reported increase in GR activity in leaves of zinc-deficient maize. Not much is known about the significance of enhancement in GR activity under zinc deficiency; but it is reported that increase in GR activity in response to xenobiotics and environmental stresses contributes to plant tolerance against these stresses (Malan et al. 1990; Lascano et al. 1998; Donahue et al. 1997).

There are several reports of zinc affecting the activities of non-zinc enzymes. Inadequate supply of zinc causes a decrease in the activity of ribulose biphosphate carboxylase (Jyung and Camp, 1976), fructose 1,6 biphosphatase (Shrotri et al. 1983) and fructose 1,6 biphosphate aldolase (O'Sulliuan, 1970; Agarwala et al. 1976; Sharma et al. 1981; Quinland and Miller, 1982; Pandey et al. 2002b). Bisht et al. (2000b) have reported a sixfold increase in alanine aminotransferase activity in leaves of tomato plants subjected to zinc deficiency. On recovery from zinc deficiency, the enzyme activity was restored to near normal.

Zinc is reported to exert a strong inhibitory effect on the plasma membrane-bound NADPH oxidase, which catalyzes the production of superoxide (O^-_2) ions. Cakmak and Marschner (1988b) reported increased activity of NADPH oxidase in roots of zinc-deficient cotton plants. Supplying zinc to the deficient plants reversed the effect. Pinton et al. (1994) reported marked enhancement in the NADPH oxidase activity, with concomitant increase in O^-_2 production in plasma membrane vesicles isolated from zinc-deficient bean plants.

Several workers have reported the enhancement of ribonuclease (RNase) activity under zinc deficiency (Dwivedi and Takkar, 1974; Johnson and Simons, 1979; Sharma et al. 1981a; Chatterjee et al. 1998; Bisht et al. 2002b; Pandey et al. 2002b). Increase in RNase activity, even before the external manifestation of zinc deficiency (symptoms), prompted Dwivedi and Takkar (1974) to suggest RNase as a biochemical parameter for diagnosing zinc deficiency. The increase in RNase activity is related to the extent of the deficiency (Sharma et al. 1981a) and is reversed on raising Zn supply from deficiency to sufficiency level (Bisht et al. 2002b). Zinc deficiency induced activation of RNase is associated with a decrease in the total RNA concentration (Sharma et al. 1981a; Bisht et al. 2002b). These observations suggest a role of zinc in regulation of RNA metabolism. Possibly, zinc prevents RNA degradation through allosteric inhibition of RNase. Low cellular concentration of zinc negates its inhibitory action on the enzyme, leading to accelerated degradation of RNA and, consequently, a decrease in protein synthesis. Similar to RNase, acid phosphatases are also activated under Zn deficiency (Hewitt and Tatham, 1960; Sharma et al. 1981a; Bisht et al. 2002b; Pandey et al. 2002b). Activation of acid phosphatases under Zn deficiency is also reversed on recovery from deficiency. Zinc is possibly involved in the maintenance of organic pool of phosphorus.

Presently, there is lack of direct evidence to substantiate zinc regulation of non-zinc enzymes in higher plants, but in yeast and animal cells, a number of metabolically critical enzymes are known to be inhibited by zinc and activated on removal of the metal from the inhibiting site (Maret et al. 1999). It is not unlikely that zinc may influence the synthesis of enzymes with no specific Zn requirement by virtue of its role as a constituent of RNA polymerase and zinc finger transcription proteins.

5.4.3 Photosynthesis

Plants exposed to zinc deficiency show decreased rate of photosynthesis (Fujiwara and Tstsumi, 1962; Ohki, 1976; Ghildiyal et al. 1978; Shrotri et al. 1981; Pandey and Sharma, 1989; Hu and Sparks, 1991; Sharma et al. 1994). Zinc nutrition effect on photosynthesis may involve changes in chloroplast structure, photosynthetic electron transport and/or CO_2 fixation. Zinc deficiency induces changes in chloroplast ultrastructure (Thomson and Weier, 1962; Shrotri et al. 1978). Leaves of zinc-deficient maize plants show abnormalities in both mesophyll and bundle sheath chloroplasts. While mesophyll chloroplasts show a decrease in the number of grana and size of the granal thylakoids, the bundle sheath chloroplasts show, increased inter lamellar spaces and presence of osmophyllic globules (Shrotri et al. 1978). Low rates of photosynthesis in zinc-deficient plants could also result from damage to the thylakoid membranes caused by enhanced generation of reactive oxygen species under zinc deficiency (Marschner and Cakmak, 1989; Cakmak and Engels, 1999). Henriques (2001) has shown that zinc deficiency causes disorganization of chloroplast thylakoids, followed by degeneration of thylakoid and stromal components, causing a corresponding decrease in the photosynthetically active area of leaves. Effect of zinc on photosynthesis may also involve inhibition of carbonic anhydrase (CA) activity (Badger and Price, 1996). In C_4 plants, CA catalyzes the production of bicarbonate (HCO^-_3), which functions as the substrate of phosphoenolpyruvate carboxylase (O'Leary, 1982; Hatch and Burnell, 1990). In C_3 plants, CA contributes to a CO_2 concentration mechanism at the site of carboxylation by RuBP carboxylase by facilitating the conversion of HCO^-_3 to CO_2 (Cooper et al. 1969). This may become crucial under zinc deficiency, when there is a decrease in CO_2 concentration because of increased resistance to CO_2 diffusion following closure of stomata (Sharma et al. 1982, 1995c; Pandey and Sharma, 1989; Sasaki et al. 1998).

Sharma et al. (1995c) have suggested CA involvement in stomatal functioning, and this may compound the effect of zinc deficiency on photosynthesis.

Srivastava et al. (1997) showed decreased photosynthetic rate, CO_2 partitioning and essential oil accumulation in leaves of zinc-deficient peppermint. Incorporation of $^{14}CO_2$ in essential oil in young developing leaves, which are actively involved in oil biosynthesis, was also decreased. The authors inferred that inadequate supply of zinc limits the production of photoassimilates, as also their availability for essential oil biosynthesis. When zinc supply is not adequately met, the products of photosynthesis are possibly utilized in primary metabolism in preference to essential oil biosynthesis.

5.4.4 Carbohydrates

Leaves of zinc-deficient plants show increased accumulation of carbohydrates (Sharma et al. 1982; Marschner and Cakmak, 1989; Marschner et al. 1996). Sharma et al. (1982) showed marked accumulation of starch in Zn-starved leaves of cabbage. Two possible explanations have been advanced for increased accumulation of starch in Zn-deficient plants. The first explanation involves limitation in shoot–root partitioning of photoassimilates (Marschner and Cakmak, 1989), possibly because of a limitation in phloem loading. The second involves a limitation in development of adequate sink capacity resulting from reduced male fertility and seed set under zinc deficiency (Sharma et al. 1990). Zinc is also suggested to be involved in starch metabolism (Jyung et al. 1975) and sucrose biosynthesis (Shrotri et al. 1980).

5.4.5 Nucleic Acids, Proteins

There are several reports of increased accumulation of non protein nitrogen and decrease in protein content in leaves of zinc-deficient plants (Sharma et al. 1982; Kitagishi and Obata, 1986; Obata and Umebayashi, 1988; Obata et al. 1999; Cakmak et al. 1989; Bisht et al. 2002b). Obata and Umebayashi (1988) showed a requirement of zinc for synthesis of protein in cultured cells of tobacco. Decrease in protein content in response to Zn deficiency has been attributed to Zn deficiency induced changes in ribosomes and nucleic acid metabolism. Leaves of zinc-deficient plants show a decrease in concentration of ribonucleic acid (Sharma et al. 1982; Bisht et al. 2002b). As a constituent of 80s ribosomes, zinc deficiency leads to a loss of integrity of ribosomes. Meristematic tissue of zinc-deficient rice plants shows a decrease in the number of 80s ribosomes associated with a decrease in protein synthesis (Kitagishi et al. 1987). Role of zinc as a constituent of transcription factors may also contribute to decreased protein synthesis.

5.4.6. Auxin Metabolism

Visible symptoms such as stunting of shoot growth and epinastic curvature of leaves and decrease in auxin concentration of zinc-deficient plants (Skoog, 1940; Tsui, 1948) suggested involvement of zinc in auxin metabolism. Singh et al. (1981) found zinc-deficient rice plants to be low in tryptophan, which functions as the precursor of auxin in one of its biosynthetic pathways. Application of tryptophan to corn grown on Zn-deficient soil was also reported to benefit seedling growth (Salami and Kenfic, 1970). These observations were taken as suggestive of a role of zinc in tryptophan dependent of biosynthesis of IAA. Cakmak et al. (1989) reported Zn deprivation of bean plants to cause to stunting of shoot growth and leaf epinasty, associated with decreased endogenous level of IAA. Resupply of Zn to Zn-deficient plants restored the level of IAA to that found in the Zn-sufficient plants. Cakmak et al. (1989) attributed the Zn effect on shoot growth to oxidative degradation of IAA. Such a situation could results from Zn-deficiency induced activation of IAA decarboxylating peroxidases, commonly called IAA oxidase. Inhibitions of the shoot growth and leaf deformities form a common response of Zn deficiency, but they are not always associated with decrease in endogenous levels of IAA. In radish, Zn-deficiency produces symptoms characteristics of Zn-deficiency with little effect on IAA concentration in the shoots (Domingo et al. 1992; Hossain et al. 1998). Hossain et al. (1998) investigated effect of zinc on the alkali-labile fractions of IAA and found that their concentration in shoots of Zn-deficient and Zn-sufficient plants was about the same. They suggested the possibility of a role of Zn in influencing the biological activity of auxin.

5.4.7 Reproductive Development and Yield

Zinc nutrient status of plants plays an important role in plant reproduction. Its deficiency inhibits different stages of plant reproductive development. Effects range from induction/initiation of flowering, floral development, male and female gametogenesis, fertilization and seed development. Even under conditions of moderate zinc deficiency, when dry matter production is only marginally depressed, the development of anthers in wheat is severely retarded (Sharma et al. 1979b, 1987, 1990). When grown with inadequate zinc or when zinc supply is withheld at the onset of tassel emergence, maize plants show reduction in the size of anthers and poor development of sporogenous tissue. Sharma et al. (1987) showed that instead of forming sporogenous tissue (pollen mother cells), about 60% anthers of the top two florets of moderately Zn-deficient maize plants developed vascular tissue (vassels). The anthers of the Zn-deficient flowers displayed repression of male sexuality. Sharma et al. (1990) showed zinc to be critical for microsporogenesis and pollen fertility. Inadequate supply of zinc leads to male sterility, impeding fertilization and seed setting.

Polar (1970) showed that pollen grains of broad bean contained a higher concentration of zinc than in other plant parts. It was shown that the pollen tubes accumulated a larger proportion of zinc in the tip and that at least a part of it played a role in fertilization. Ender et al. (1983) showed higher accumulation of zinc in the growing tip of lily pollen than its basal parts. In field bean (Pandey et al. 1995b) and green gram (Pandey et al. 2000), zinc deficiency has been shown to induce changes in exine morphology and reduce pollen viability. The exine of Zn-deficient pollen grains lacks the reticulate and uniform muri present in the exine of the Zn-sufficient pollen grains. Instead, Zn-deficient pollen grains show highly sinuous, lobed muri having a waxy covering and raised baculae (Fig.5.2). Changes in exine pattern are reported to be caused by corresponding changes in mRNA and protein levels in pollen grains (Willing et al. 1988; Wetzel and Jensen, 1992). Zinc deficiency is reported to decrease the RNA content of pollen grains, possibly because of elevated levels of RNase activity (Sharma et al. 197b; 1987). This may possibly lead to alterations in the exine pattern of Zn-deficient pollen grains. Zinc deficiency has also been shown to induce changes in stigmatic size, morphology and exudations,

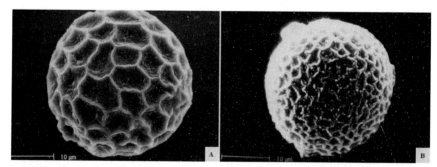

Fig. 5.2 Scanning electron micrograph showing the effect of Zn deficiency on pollen size and exine architecture in green gram (*Vigna radiata* L cv T-44). The Zn-sufficient pollen grains (a) show reticulate ornamentation with uniform muri. In Zn-deficient pollen (B), reticulation is incomplete and muri are highly sinous and lobed. Muri also show wax deposition and prominent raised baculae in the lumen (After Pandey and Sharma. 2000).

inhibiting pollen-stigma interaction (Pandey et al. 1995b, 2005). Scanning electron microscopic (SEM) examination of the stigmatic surface of Zn-deficient field bean flowers revealed a decrease in the pollen receptive area of the stigmatic head.

Floral aberrations and inhibition of microsporogenesis in zinc-deficient plants may involve the role of zinc as a constituent of zinc finger proteins. The TF IIIA-type zinc finger proteins have been shown to control the development of floral parts by controlling cell division and/or expansion

Fig. 5.3. Scanning electron micrograph showing Zn-deficiency effect on stigmatic head of faba bean (*Vicia faba* L) and number of pollen adhering to it. Compared to Zn-sufficient stigma (A), only a few pollen grains are seen on the surface of Zn-deficiency stigma (B) which lacks in stigmatic exudations because of unruptured cuticle covering the papillae. (After Pandey et al. 1995).

of particular cell types in petunia (Takasuji et al. 1992) and *Arabidopsis* (Sakai et al. 1995). The anther-specific zinc finger proteins are suggested to play a role in microsporogenesis (Kobayashi et al. 1998) and the zinc finger polycomb proteins are suggested to be involved in megagametogenesis and seed development (Grossniklaus et al. 1998; Brive et al. 2001). Reproductive development, including sporogenesis, could be impaired because of limited delivery of zinc to the reproductive organs. In a recent study, Takahashi et al. (2003) have shown severe suppression of reproductive development in transgenic tobacco plants (naas transgenics) lacking in synthesis of nicotianamine, which plays a role in the chelation of the cationic micronutrients and their delivery to the reproductive tissues.

Seeds of zinc-deficient plants take a longer time to mature (Boawn et al. 1969; Baylock, 1995; Morghan and Grafton, 1999). Sharma et al. (1995b) reported alterations in seed coat topography, which could possibly inhibit their germination. Based on the study of Zn levels at different stages of seed development in *Arabidopsis*, Otegui et al. (2002) have reported accumulation of Zn, as Zn-phytate, in the chalazal vacuoles of the endosperm and its disappearance at the late globular stage. Expression of most seed proteins during embryogenesis was taken to suggest that Zn-phytate stored in the chalazal vacuoles of the endosperm functions as the Zn source for the assembly of several diverse Zn–proteins, when their need arises during the course of embryo development (Otegui et al. 2002).

5.4.8 Water Relations

Suboptimal supply of zinc in sand culture induces changes in plant water relations parameters and leaf morphology, similar to those observed in response to water deficit. Sharma et al. (1982) showed that leaves of cabbage plants supplied with Zn-deficient nutrition turned thick and leathery and the leaf epidermis developed a thick waxy coating. The water content of the Zn-deficient leaves per unit leaf area was high, yet they showed higher water saturation deficit (WSD) and more negative water potential than corresponding leaves of Zn-sufficient plants. Similar observations were made for cauliflower. Limitation in zinc supply to cauliflower caused drastic reduction in stomatal opening, associated with increase in diffusive resistance and decrease in transpiration rate (Sharma and Sharma, 1987). Pandey and Sharma (2000) have described micro morphological changes in leaf epidermis of faba bean (*Vicia faba*) grown at low (deficient) level of Zn supply (Fig. 5.4). The guard cells and the accessory cells showed concentric rings or folds (which became more prominent on removal of epicuticular wax); the stomatal lips were distorted and stomatal aperture was reduced (Sharma et al. 1995c; Pandey and Sharma, 2000).

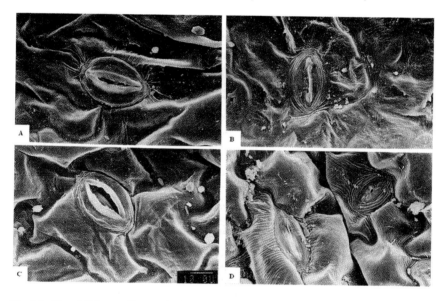

Fig 5.4. Zinc deficiency effect on leaf micro morphology of faba bean (*Vicia faba* L. cv. VH-130). In contrast to Zn-sufficient leaves (A), leaves of Zn-deficient plants (B) show thick epicuticular wax, flaccid guard cells and closed stomata. Removal of the wax coating with chloroform shows that stomata of Zn-sufficient leaves (C) are wide open but those of Zn-deficient leaves (D) are sunken and closed, with their guard and accessory cells having concentric wrinkles. (After Sharma et al. 1995c).

Increased accumulation of proline, a characteristic feature of water-stressed plants, is also a common feature of zinc-deficient plants (Ghildiyal et al. 1977; Sharma et al. 1982, 1984a, 1995c; Sharma and Sharma, 1987). Induction of changes, characteristic of water deficit, in zinc deficient but well-watered plants, suggests development of physiological water stress, possibly involving a common target or signaling mechanism under zinc deficiency, and water deficit.

5.4.9 Oxidative Stress

Zinc deficiency accelerates oxidative stress both by overproduction of free oxygen radicles and restricted detoxification of the reactive oxygen species (Cakmak, 2000; Obata et al. 2001). Zinc deficiency causes activation of the membrane bound NADH-oxidase leading to enhanced production of superoxide ions (Cakmak and Marschner, 1988b). It also decreases the activities of superoxide dismutase (Cu-Zn SOD), a key enzyme of the antioxidant defence system (Cakmak et al. 1997c; Yu and Rengel, 1999; Obata et al. 1999; Pandey et al. 2002b, c). Decrease in Cu-Zn SOD activity not only leads to toxic build up of superoxide ions (O_2^-) but also their conversion to even more toxic OH^- radicles involving Fenton chemistry. The hydroxyl radicles thus generated cause oxidative damage to different cellular constituents, more particularly the membrane lipids. Figure 5.5 shows some Zn-deficiency induced changes in antioxidant defense system of maize resulting in enhanced lipid peroxidation (Pandey et al. 2002c). Plants that are inadequately supplied with zinc show enhanced damage from oxidative stress imposed by photoinhibitory conditions or ozone injury (Wenzel and Mehlorn, 1994).

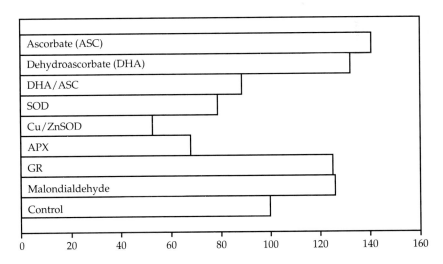

Fig 5.5. Relative values (-Zn/+Zn×100) of antioxidants, antioxidant enzymes and malondialdehyde in leaves of zinc deficient maize (*Zea mays* L.) plants. SOD-superoxide dismutase; APX-ascorbate peroxidase; GR-glutathione reductase (After Pandey et al. 2002).

Enhanced peroxidation of membrane lipids cause structural changes in plasma membranes (Van Ginkel and Sevanian, 1994) that makes them more leaky (Welch and Norvell, 1993). Welch et al. (1982) had reported leakage of salts by roots of Zn-deficient plants. Salt incrustations on the leaf surfaces of Zn-deficient plants (personal observations) also suggest leakage of salts through the leaf epidermis. Enhanced peroxidation of lipids of the thylakoid membranes may result in major damage to the photosynthetic apparatus (Marschner and Cakmak, 1989; Cakmak and Engels, 1999; Henriques, 2001) and lower the photosynthetic efficiency of plants.

CHAPTER 6

MOLYBDENUM

6.1 GENERAL

Molybdenum is a metal of the second transition series. It is a subgroup VI B element. The 4 d electrons in the outermost shell enable it to exist in several oxidation states—Mo(III), Mo(IV), Mo(V), Mo(VI)). The most stable oxidation state is the hexavalent [Mo(VI)] form. Molybdenum has high affinity for oxygen. Different oxidation states form different oxycations, which in an aqueous system, forms coordination complexes with water molecules. Coordination of two molecules of water with the oxycation $Mo(VI)O_2^{2+}$ forms strongly acidic protons which dissociate the cation to form the molybdate anion-Mo(VI), which constitutes the predominant form of molybdenum in plants. In this form [Mo(VI)], the chemical properties of molybdenum have many similarities with vanadium and tungsten (which in some functions partially replace molybdenum). Molybdenum also has an affinity for sulphur-containing groups. Easy convertibility of hexavalent form [Mo(VI)] to the reduced—[Mo(II), Mo(IV)]—forms enables it to function in oxidation-reduction reactions. Singly, or in combination with other essential elements (Fe, Cu, S), molybdenum functions as a cofactor of several enzymes (Hille, 1996; Sigel and Sigel, 2002). The role of molybdenum in assimilation of nitrate and dinitrogen fixation involves its functioning in oxidation-reductions involving electron transport. Molybdenum is essential for nitrogen fixation by free living as well as symbiotic bacteria.

The chemical properties of the molybdate ion resemble those of sulphate and phosphate, which accounts for interactive effects of Mo, S and P in their assimilation and metabolism.

6.2. UPTAKE, TRANSPORT

Molybdenum is taken up as the molybdate anion (MoO_4^{2-}). It has been suggested that uptake of molybdenum involves the same anion transporter as is involved in the transport of phosphate (PO^-_4) (Heuwinkel et al. 1992). When phosphate supply was limiting (deficient), tomato plants showed as much as fivefold increase in uptake of radio-labelled MoO^-_4. Inhibition of molybdate uptake by large amounts of sulphate (SO_4^{2-}) (Marschner, 1995) also suggested that transport of MoO_4^{2-} involves a common anion transporter. Mutant analysis of *Chlamydomonas reinhardti* (Llmas, 2000) suggests involvement of more than one transporter in molybdenum uptake. Molybdenum uptake by *Chlamydomonas* involves two transporter systems, a 'high-affinity, low-capacity transporter', which is inhibited by 0.3 mM SO_4^{2-}, and a 'low-affinity, high-capacity transporter', that is not inhibited by SO_4^{2-}. While the former is insensitive to tungstate, the latter is inhibited by it. Bulk transport of molybdenum has, however, been suggested to take place through the low-affinity system described for tomato by Heuwinkel et al. (1992).

Information on the form in which molybdenum is transported from roots to shoots is limited. On the basis of electrophoretic mobility of molybdenum in xylem sap, Tiffin (1972) suggested xylem transport of molybdenum in the ionic (MoO_4^{2-}) form or as molybdate complexes of sulphur amino acids, sugars or polyhydroxy compounds. Early experiments by Bokovac and Wittwer (1957) suggested that phloem mobility of molybdenum figures in between the highly mobile elements such as phosphorus and the phloem immobile elements such as calcium. Jongruaysup et al. (1994) showed that in black gram phloem mobility of molybdenum varied with the level of molybdenum supply. It was phloem mobile in all parts of molybdenum-sufficient plants, except possibly in stem segments, but was phloem immobile in molybdenum-deficient plants.

6.3 ROLE IN PLANTS

Molybdenum is a cofactor of 30 or more bacterial and plant enzymes catalyzing redox reactions (Hille, 1996; Stiefel, 1996). In higher plants, molybdenum enzymes nitrate reductase and xanthine oxidase catalyze reactions involving nitrogen and nitrogenous compounds, aldehyde oxidase catalyzes the terminal reaction of the biosynthetic pathway of abscissic acid (Walker-Simmons et al. 1989; Taylor, 1991) and auxins (Koshiba et al. 1996), and sulphite oxidase catalyzes oxidation of SO_3^{2-} to SO_4^{2-}.

A characteristic feature of the molybdenum enzymes (other than nitrogenase) is that they contain additional cofactors such as iron, Fe-S clusters, haem, or flavin (FAD), which aid in binding of the substrate to the enzyme and facilitate the transfer of electrons (Mendel, 1997; Mendel and Hansch, 2002). The molybdenum cofactor (MoCo) of the molybdenum enzymes (other than nitrogenase) is essentially the same. This was first suggested by Pateman et al. (1964) on the basis of their analysis of

Aspergillus nidulans mutants. Mutants that lacked in nitrate reductase also lacked in xanthine dehydrogenase (Mendel and Miller, 1976). Subsequently, it was shown that addition of denatured preparations of purified molybdenum enzymes, other than nitrate reductase (NR), to crude extracts of NR deficient mutant of *Neurospora crassa* (*nit-1*) led to reconstitution of NR activity (Nason et al. 1971).

Rajagopalan and Johnson (1992, references therein) showed that the organic moiety to which Mo binds to form the MoCo is a pterin. Unlike the other known pterins, this pterin has a tricyclic structure with a four-carbon alkyl side chain containing a dithiolene group and a terminal phosphate ester (Fig 6.1). Molybdenum is bound to the dithiolene group of pterin. The molybdenum containing pterin, referred to as **Molybdopterin** (MPT), is an essential component of the molybdenum cofactor (MoCo). Crystallographic analysis of Mo-enzymes provided further evidence of binding of molybdenum to pterin and showed that the cofactor is embedded deep within the apoprotein and that the pterin ring structure is involved in the transfer of electrons to or from the Mo atom (Kisker et al. 1997; Campbell, 1999).

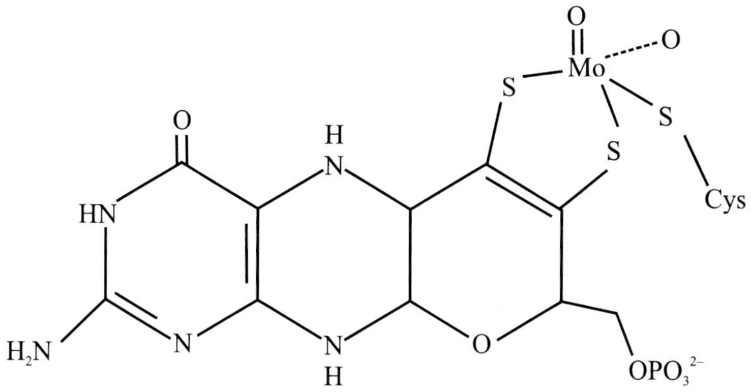

Fig. 6.1. Structure of Mo-molybdopterin (Mo-MPT) of nitrate reductase Molybdenum is shown coordinated to the pterin through its dithlolane group (After Campbell, 1999)

Different aspects of the molybdenum enzymes, their cofactors, biosynthesis and molecular biology are covered by Mendel and associates (Mendel, 1997; Mendel and Schwartz, 1999; Mendel and Hansch, 2002).

NITRATE REDUCTASE (EC 1.6.6.1-3)

Nitrate reductase catalyzes the reduction of nitrate to nitrite. The bacterial enzyme (respiratory or dissimilatory nitrate reductase) catalyzes the reduction of nitrate in denitrifying bacteria. The eukaryotic enzyme, assimilatory nitrate reductase, catalyzes the first step in the reduction of nitrate to ammonia ($NO_3^- \rightarrow NO_2^-$). Ammonia is the form in which inorganic nitrogen is assimilated into organic nitrogenous compounds. The bacterial and the eukaryotic enzymes belong to two distinct families of molybdoenzymes.

The assimilatory nitrate reductase of higher plants is a large homodimer (containing about 1,000 amino acids). Fragmentation of the holoenzyme, following partial proteolysis and measurement of the enzyme activities associated with the different fragments, showed each subunit to contain three cofactors—a molybdenum cofactor (MoCo), a b-type cytochrome (Haem-Fe) and a flavin nucleotide (FAD) (Fig. 6.2). The three cofactors are spatially arranged so that they form three distinct structural regions, each functioning as a separate redox centre catalyzing a specific electron transport reaction. (Campbell and Kinghorn, 1990; Garde et al. 1995; Ratnam et al. 1995; George et al. 1999).

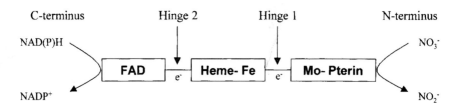

Fig 6.2. Schematic model of eukaryotic nitrate reductase (NR). The enzyme has three functional domains: a flavin adenosine dinucleotide (FAD) cofactor near the C-terminal, a haem Fe cofactor in the middle and a Mo-pterin cofactor near the N-terminal. The haem Fe is separated from the other two cofactors by hinge regions. (After Campbell, 2001).

Structure of the enzyme based on cDNA sequences revealed that the Mo cofactor domain is located near the N terminus, the haem-Fe domain in the middle and the FAD domain at the C terminus, with the three functional domains being connected by two hinges (H_1 and H_2) (Campbell, 1999). In the initial reaction, the FAD cofactor accepts two electrons from NADH or NADPH. Most forms of NR use NADH but some can use either (NAD(P)H). The reduced FAD transfers the electrons to the haem-Fe, which shuttles them to the Mo cofactor. The Mo cofactor donates the electrons to nitrate, reducing it to nitrite. Transport of electrons from NAD(P)H to nitrate through the three cofactors is facilitated by the difference in the mid point potential (E'^o) of the three cofactor domains and the light/dark reversible regulation of NR is attributed to phosphorylation of the Ser residue in hinge 1 (Campbell, 2001). The overall reaction catalyzed by nitrate reductase can be shown as:

$$NO_3^- + NAD(P)H + H^+ \rightarrow NO_2^- + NAD(P)^+ + H_2O$$

Within the plant distribution of nitrate reductase differs in different plant spp but in all the plants the enzyme is localized to the cytosol. In maize (C_4), NR is localized in the mesophyll cells. The activity of nitrate reductase is regulated by nitrate concentration and several other factors such as light, CO_2, cytokinins and concentration of some cellular metabolites like glutamine and sucrose. Enzyme activity is regulated at both transcriptional (gene expression) and post-translational levels. The former contributes to long-time responses induced by NO_3^-, glutamine, sucrose, cytokinins and nitrogen starvation and the latter to rapid responses produced by O_2 and CO_2 concentrations and by light.

Nitrate reductase is also known to catalyze one electron transfer from NAD(P)H to molecular oxygen producing superoxide (O_2^-) ion (Barber and Kay, 1996) and to nitrite-producing nitric oxide (Dean and Harper, 1988; Yamasaki et al. 1999; Yamasaki and Shakihama, 2000; Rockel et al. 2002; Kaiser et al. 2002). Desikan et al. (2002) presented genetic evidence of NR- mediated ABA-induced NO systhesis in guard cells of *Arabidopsis thaliana*. These observations point to as yet little understood role of NR in generation of reactive species of oxygen and nitrogen.

XANTHINE DEHYDROGENASE (EC 1.1.1.204)

Xanthine dehydrogenase (XDH) is a homodimeric molybdoprotein (Montalbini, 1998). It has two identical and catalytically independent subunits. As in case of nitrate reductase, each subunit contains a molybdopterin, two iron-sulphur (Fe_2-S_2) clusters and one flavin (FAD) cofactor. The enzyme catalyzes the two-step oxidative degradation of xanthine and hypoxanthine to uric acid using NAD^+ as an electron acceptor. The animal enzyme can be converted to xanthine oxidase (EC 1.1.3.22.)

$$Xanthine + O_2 + H_2O \xrightarrow{e^-} Uric\ acid + H_2O_2$$

Xanthine-oxidizing capacity has been reported from several plant sources (Mendel and Muller, 1976; Nguyen and Feierabend 1978; Nguyen, 1986). Xanthine dehydrogenase (XDH) from wheat (Montalbini, 1998) and pea (Sauer et al. 2002) have been purified and characterized. XDH is involved in purine catabolism and ureide synthesis (Schubert and Boland, 1990). It plays a role in the nitrogen metabolism of legumes, in which nitrogen fixed in the root nodules is transported to the host (soybean, cowpea) in the form of ureides (allantoin and allantoic acid). Decrease in the activity of XDH due to inadequate supply of molybdenum may impair the oxidative breakdown of purines and lead to their accumulation in the root nodules of the 'ureide type' plants. XDH has also been suggested to be involved in generation of superoxide radicals and plant-pathogen relationships (Montalbini, 1992). Senescing leaves of pea are reported to show high activity of XDH associated with enhanced generation of reactive oxygen species (Pastori and Rio, 1997). XDH has been suggested to play a role in regulation of iron homeostasis (Navarre et al. 2000) and production of nitric oxide (Harisson, 2002), which functions as a signaling molecule (Neill et al. 2003).

Sulphite Oxidase (EC 1.8.3.1)

Essentially a human liver enzyme, sulphite oxidase (SO) is also reported from plants (Jolivet et al. 1995; Eilers et al. 2001) SO is a dimer, but unlike the other molybdenum oxido-reductases, lacks the redox-active centres other than MoCo (Mendel and Hansch, 2002). SO catalyzes the oxidation of sulphite to sulphate:

$$SO_3^{2-} + H_2O \longrightarrow SO_4^{2-} + 2e^- + 2H^+$$

The main role of sulphite oxidase is suggested to be detoxification of sulphite, which may accumulate in the peroxisomes in toxic concentrations. Even low levels of sulphite are reported to be inhibitory to catalase (Veljovic – Jovanovic et al. 1998).

Aldehyde Oxidase (EC 1.2.3.1.)

Aldehyde oxidase (AO), like nitrate reductase, contains FAD, iron and MoCo as prosthetic groups (Koshiba et al. 1996). AO has broad substrate specificity (Sekimoto et al. 1998). It catalyzes the oxidation of abscisic aldehyde to abscissic acid (ABA) (Seo and Koshiba, 2002) and indole-3-acetaldehyde to indole-3-acetic acid (IAA) (Koshiba et al. 1996; Seo et al. 1998). Both ABA and IAA play major regulatory roles in developmental processes of plants. Broad substrate specificity of AO also enables it to function in many other metabolic reactions including detoxification reactions and response to pathogenic attack (Mendel and Hänsch, 2002).

Nitrogenase (EC 1.18.6.1)

The prokaryotic enzyme nitrogenase, which catalyzes the fixation of atmospheric nitrogen (N_2), is a complex of two enzymes—**dinitrogenase** (the **Mo-Fe protein**) and **dinitrogenase reductase** (also known as **Fe-protein**). Dinitrogenase is an $\alpha_2\beta_2$ heterotetramer containing an iron-molybdenum cofactor (FeMoCo). Under molybdenum deficiency, nitrogenase of azotrophic bacteria may have a Fe or a FeMo cofactor, but the symbiotic bacteria have an absolute requirement for FeMoCo.

Dinitrogenase reductase is a homodimer containing a single 4Fe-4S cluster that cooperatively holds the two monomers to the protein surface. It provides high reducing power to dinitrogenase. The presence of nitrogenase complex in diazotrophic bacteria and in root nodules of higher plants, capable of symbiotic nitrogen fixation enables biological reduction of nitrogen, which is chemically formidable, because the N≡N bond is extremely strong (bond energy-225 kCal mol^{-1}) and resistant to chemical action. Biological fixation of nitrogen requires an input of eight high

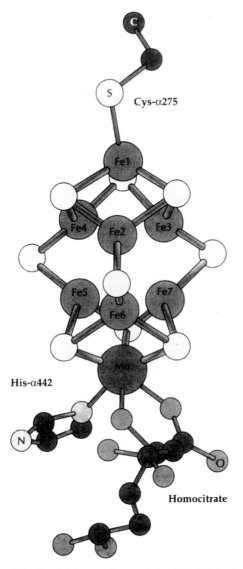

Fig 6.3. Molecular model of molybdenum-iron cofactor (FeMoCo) of symbiotic bacterial nitrogenase. The (FeMoCo) consists of one (4Fe-3S) cluster and one (1Mo-3Fe-3S) cluster bridged by three atoms of inorganic sulphur. The Mo atom is linked to homocitrate. (Reprinted, with permission, from *Biochemistry and Molecular Biology of Plants.* BB Buchnam, W Gruissem, RL Jones, eds ©2000 American Society of Plant Biologists).

potential electrons and eight protons for catalyzing the reduction of one molecule of nitrogen to ammonia ($N_2 \rightarrow NH_3$).

$$N_2 + 16 \text{ ATP} + 8e^- + 8H^+ \rightarrow 2NH_3 + H_2O + 16 \text{ ADP} + 16 \text{ Pi}$$

Nitrogenase is highly sensitive to inactivation by O_2. In leguminous plants, which fix nitrogen symbiotically, the enzyme is prevented from inactivation by the iron protein leghaemoglobin. Leghaemoglobin has very high affinity for O_2 and restricts the root nodule concentration of O_2 to levels that are sufficient to meet the growth requirement of bacteroids but not inhibitory to nitrogenase.

6.4 DEFICIENCY RESPONSES

6.4.1 Visible Symptoms

Symptoms of molybdenum deficiency in a wide variety of plant have been described by Hewitt (1956) and Gupta (1997d). In general, symptoms appear late in the growing season and initiates as fine chlorotic mottling of old or lower middle leaves. Plants belonging to *Brassicaceae* are particularly sensitive to molybdenum deficiency. In cauliflower, molybdenum deficiency symptoms appear in the form of chlorotic mottling and cupping of middle leaves. Leaf margins appear scorched and withered. Scorching and tissue disintegration gradually extends to the entire lamina. If deficiency persists, the young emerging leaves fail to expand and apical growth is arrested. Under moderate deficiency, after apparently normal growth for some time, the middle leaves develop small patches in between the veins. The lamina underlying these patches gradually turns translucent and disintegrates. Eventually, the leaves are left with no lamina, except in a little strip along the midrib (whiptail syndrome). Chloroplast disintegration precedes the loss of lamina and is attributed to enhanced peroxidation of chloroplast membrane lipids (Fido et al. 1977). The damage does not involve any direct role of molybdenum and can, to a large extent, be prevented by substitution of tungsten for molybdenum. In radish, which is also sensitive to molybdenum deficiency (pers. obs.), old leaves show yellow-green mottling in between the veins. Leaf margins appear to be wilted. The entire lamina appears bleached and curls downward. Somewhat similar symptoms are exhibited by tomato. Old and middle leaves of molybdenum-deficient tomato plants show fine yellow-green mottling with their margins curled inwards. In beans, and other broad leaf legumes (e.g. cowpea), which are also sensitive to molybdenum deficiency, old leaves of molybdenum-deficient plants develop chlorotic mottling in between veins and this is followed by necrosis of the intervenal areas, commonly referred to as 'scald'. *Citrus* shows somewhat different but characteristic symptoms (Yellow spot) in response to molybdenum deficiency. Leaves appearing during the early-summer flush develop water-soaked areas which later enlarge, turn yellow and coalesce to form linear or irregular patches except along a narrow area adjacent to the veins. Leaves affected thus fall prematurely causing denudation of branches.

Cereals are comparatively resistant to molybdenum deficiency. Symptoms of molybdenum deficiency appear considerably late in the growing season. In wheat plants subjected to severe deficiency of molybdenum, at the time of ear emergence, the tips of middle leaves turn golden yellow (Agarwala and Sharma, 1979). Yellowing spreads down towards the base along the leaf margins. Middle leaves may also develop small faded or chlorotic areas in the middle. These areas gradually spread to the entire leaf blade, which eventually turns dry and papery. At this stage, the tips of the young leaves twist to form spirals.

6.4.2 Enzyme Activities

Of the different enzymes, activities of which are influenced by plant molybdenum status, nitrate reductase figures most prominently. Nitrate reductase is an inducible enzyme induced by molybdenum in the presence of nitrate. Plants supplied with a reduced form of nitrogen (NH_4^+) show very low or totally absent NR activity, even when supplied with adequate molybdenum. Molybdenum deprivation of nitrate-grown plants leads to a marked decrease in NR activity (Hewitt et al. 1957; Afridi and Hewitt, 1964; Hewitt and Gundry, 1970; Agarwala et al. 1978b, 1988; Chatterjee et al. 1985). Resupply of molybdenum to molybdenum-deficient plants leads to reconstitution of NR activity (Chatterjee et al. 1985). Witt and Jungk (1977) showed a close correspondence between molybdenum supply, molybdenum accumulation and nitrate reductase activity in spinach leaves. They also showed that within 2 h of infiltration of molybdenum in leaf segments of plants subjected to molybdenum deficiency (< 0.01 mg Mo l^{-1}), NR activity was raised to about the level obtained in molybdenum-sufficient plants. Decrease in nitrate reductase activity under molybdenum deficiency and its reversal on resupply of molybdenum has also been reported by Agarwala et al. (1979, 1988). Sensitivity of nitrate reductase activity to changes in tissue molybdenum concentration has formed the basis for its use as a biochemical parameter for assessing molybdenum status of plants (Shabed and Bar-Akiva, 1967).

Besides molybdoenzymes, several other enzymes are also influenced by plant molybdenum status. Molybdenum-deficient plants show decrease in the activities of the iron enzymes—catalase (Agarwala et al. 1978b, 1986), cytochrome oxidase (Chatterjee et al. 1985) and succinate dehydrogenase (Agarwala et al. 1986). Molybdenum effect on peroxidase is, however, inconsistent. Molybdenum deficiency effect on catalase can be reversed on resumption of adequate molybdenum supply (Agarwala et al. 1978b, 1988). Chatterjee and Agarwala (1979) found decreased activities of catalase and succinic dehydrogenase in molybdenum-deficient tomato plants to be associated with reduced absorption and root to shoot transport of iron (^{59}Fe) and suggested it to be a possible outcome of molybdenum effect on availability of iron. In an earlier report, Berry and Reisenaur (1967) had shown a decrease in iron absorption by molybdenum-deficient tomato plants and attributed it to decrease in iron reduction capacity of roots. Interaction of molybdenum and iron in their uptake mechanism seems to be more complex and a promising area for future investigations.

Agarwala et al. (1978b) reported decreased activities of aldolase and alanine aminotransferase in corn plants subjected to molybdenum deficiency. Restoration of adequate molybdenum supply to low molybdenum plants led to partial restoration of the enzyme activities. Kaplan and Lips (1984) reported a decrease in the activities of glycolate dehydrogenase and glycolate oxidase under molybdenum deficiency. In an early study, Fujiwara and Tstusumi (1955) reported a decrease in the IAA oxidase activity in molybdenum-deficient plants. Activities of several enzymes are increased by molybdenum deficiency. Molybdenum-deficient plants show several-fold increase in the activity of acid phosphatase (Spencer, 1954; Hewitt and Tatham, 1960, Agarwala et al. 1978b, 1986, 1988; Chatterjee et al. 1985). This has been attributed to an inhibitory effect of molybdenum on the enzyme activity (Spencer, 1954). Molybdenum deficient plants also show enhancement of ribonuclease activity (Agarwala et al. 1978b; Chatterjee et al. 1985).

6.4.3 Nitrogen Fixation

As a constituent of nitrogenase, molybdenum plays a key role in the fixation of atmospheric nitrogen by free-living N_2 fixing microorganisms and symbiotic bacteria (*Rhizobacterium leguminosarum*) in the root nodules of leguminous and some non-leguminous plants (e.g. *Alnus glutinosa*). The leguminous plants are sensitive to molybdenum nutrition because of a dual requirement of molybdenum: one for nodulation and nodule development, and the other for the synthesis of nitrogenase. Molybdenum-deficient legumes produce many small nodules with low N_2 fixing activity. The nodules also lack the characteristic pink colour attributed to leghaemoglobin and the bacteroids show structural deformities. It has been suggested that under conditions of molybdenum deficiency, Mo requirement for root nodules is met in preference to that for the growth of the host plant (Ishizuka, 1982). Application of molybdenum to nodulating soybeans increases the size and dry weight of root nodules, particularly during later stages of plant growth (Hashimoto and Yamasaki, 1976). In a study with nodulating soybeans, Parker and Harris (1977) showed an increase in the ratio of nodule dry weight to leaf Mo content in response to Mo fertilization, which implies greater requirement of Mo for N_2 fixation than for nitrate assimilation. Efficiency of molybdenum accumulation also affects the efficiency of nitrogen fixation. Brodrick and Giller (1991) reported high rates of N_2 fixation by molybdenum-efficient genotypes of bean (*Phaseolus vulgaris*). Use of high molybdenum seeds benefits both plant growth and N_2 fixtion (Brodrick et al. 1992).

6.4.4 Carbohydrates

As a part of studies on the role of molybdenum in growth and metabolism of plants and their dependence on the source and level of nitrogen supply, Hewitt and coworkers (see Hewitt, 1963) described molybdenum deficiency induced changes in several organic compounds. Suboptimal supply of molybdenum was reported to decrease the concentration of sugars in leaves of cauliflower (Agarwala and Hewitt, 1955). Its effect on reducing sugars was particularly marked.

6.4.5 Nitrogenous Compounds, Nucleic Acids

Since nitrate reductase is an inducible enzyme, induced by molybdenum in presence of nitrate, molybdenum deficiency effect on nitrogenous compounds is influenced by the source of nitrogen. As would be expected, the nitrate-grown plants respond to molybdenum deficiency by showing enhanced accumulation of nitrate associated with decreased concentration of nitrite and organic nitrogenous compounds. Early studies with cauliflower (Hewitt et al. 1957) showed marked accumulation of nitrate, associated with a decrease in the protein and total organic nitrogen contents, in nitrate-grown plants deprived of molybdenum. The concentration of aminoacids, in particular glutamic acid, aspartic acid and alanine, especially glutamic acid, was severely reduced. Plants-supplied nitrogen as ammonium compounds showed no such effect. Instead, these plants showed high accumulation of arginine. Gruhn (1961) also reported enhanced accumulation of amino acids in molybdenum-deficient plants.

In nodulated leguminous plants, wherein N_2 is reduced to NH_3 under the catalytic influence of nitrogenase, glutamic acid forms the first organic compound into which ammonia is incorporated. This explains the observed decrease in the glutamic acid content of legumes subjected to molybdenum deficiency. In the ureide type of nodulating legumes (soybean, cowpea), low molybdenum availability in soils may inhibit transport of the ureides (allantoin and allantoic acid) to the host cells. This is in consonance with the role of Mo as a constituent of xanthine dehydrogenase. Decrease in xanthine dehydrogenase activity leads to increased accumulation of xanthine in root nodules of molybdenum-deficient plants.

Chatterjee et al. (1985) reported a decrease in the concentration of nucleic acids (DNA, RNA) in leaves of mustard (*Brassica campestris* var. Sarson) plants subjected to molybdenum deficiency. The decrease in RNA content was particularly marked. This could result from an increase in ribonuclease activity under molybdenum deficiency (Agarwala et al. 1978b; Chatterjee et al. 1985). Studies on molybdenum nutrition of cauliflower, grown with different forms of nitrogen (Hewitt et al. 1957), showed a decrease in the adenine content of nitrate-grown plants subjected to molybdenum deficiency. Plants that received nitrogen in the form of ammonium compounds showed no such effect.

6.4.6 Organic Acids

Molybdenum-deficient plants, which accumulate nitrate in high concentrations, are generally low in organic anions. Merkel et al. (1975) reported a marked decrease in malate and citrate contents under molybdenum deficiency. Compared to nitrate-grown plants, accumulation of the organic anions is larger in plants that are also supplied ammonium nitrogen. Opposite observations were, however, made by Hofner and Grieb (1979) in molybdenum-deficient plants.

6.4.7 Ascorbate Metabolism

Involvement of molybdenum in ascorbate metabolism of plants was first demonstrated by Hewitt et al. (1950). It was shown that in nine different plant species grown in sand culture, molybdenum deficiency led to a decrease in ascorbic acid content ranging from 55 to 75%. Injection of molybdenum through the petiole caused increase in ascorbic acid content, that could be detected within 24 h and attained 75 to 95% the concentration in molybdenum-sufficient plants in three to five days. The decrease in ascorbic acid content under molybdenum deficiency was found with all forms of nitrogen supply, but was most pronounced with nitrate (Agarwala and Hewitt, 1954; Hewitt and McCredy, 1956). Agarwala and Hewitt (1954) also observed an increase in the dehydroascorbic acid/ascorbic acid ratio in molybdenum-deficient plants.

6.4.8 Reproductive Development, Seed Physiology

Sand culture studies on molybdenum nutrition of maize (Agarwala et al. 1978b, 1979), sorghum (Agarwala et al. 1988) and wheat (Chatterjee and Nautiyal, 2001) have shown that reproductive development, including grain yield, is sensitive to changes in molybdenum nutrition. Deprivation of adequate supply of molybdenum to corn leads to severe retardation in cob size, failure of styles (silk) to protrude out of the husk, and poor setting of grains (Agarwala et al. 1978b). Molybdenum deficiency effect on development of tassels and pollen grains in maize is marked (Agarwala et al. 1979). The development of tassels and their emergence is delayed and the lemma and the palea turn severely chlorotic and papery. Under severe deficiency (0.01 μg Mo l^{-1}), anther size is severely restricted and the anthers appear shriveled (because of lack of sporogenous tissue). Even under less severe deficiency (0.1 μg Mo l^{-1}), the pollen-producing capacity is severely limited, and the size of the pollen grains is reduced (Table 6.1). Pollen grains lack dense cytoplasmic contents and fail to stain with iodine-potassium iodide solution, suggesting lack of starch and poor viability. The pollen grains also show low activity of invertase, which reflects poor sucrose utilizing capacity and loss of viability. This is substantiated by poor *in vitro* germination of pollen grains of molybdenum-deficient plants.

Table 6.1. Molybdenum effect on some pollen attributes of maize (*Zea mays* L.)

	Molybdenum supply (μg l^{-1})		
	0.01	0.1	20
Mo concentration in pollen grain (μg g^{-1} dry wt.)	0.017	0.061	0.092
Pollen grains anther^{-1} (no.)	1,300	1,937	2,437
Pollen diameter (μm)	68	85	94
Pollen germination (%)	27	51	86

(After Agarwala et al. 1979)

Gubler et al. (1982) reported a phenomenal increase in the number of melon (*Cucumis melo* L.) fruits produced on acid soil in response to molybdenum application. This could possibly be due to a beneficial effect of molybdenum on pollen fertility. Chatterjee and Nautiyal (2001) have described molybdenum effect on development of wheat seeds, their chemical composition and vigour. Seeds produced on molybdenum-deficient plants show decreased endosperm reserves. The concentrations of starch, sugars, protein and non-protein nitrogen are also decreased. Seed protein also shows qualitative changes. Relative to seeds of Mo-sufficient plants, seeds of Mo-deficient plants are low in prolamins, glutelin and globulin and rich in albumin. In chickpea also, molybdenum deficiency leads to qualitative changes in seed proteins. While the relative concentration of legumins and vicilins is reduced, that of albumins is increased (Nautiyal et al. 2005). Chatterjee and Nautiyal (2001) had considered the increase in the relative proportion of albumins in seeds of molybdenum-deficient wheat to be a factor contributing to their poor viability and loss of vigour.

Deficiency of molybdenum reduces seed dormancy and causes pre-harvest sprouting of cereals grains. Low molybdenum seeds of maize are reported to sprout on the cobs (Tanner, 1978; Farwell et al. 1991). In wheat, premature sprouting of grains can be controlled by foliar application of molybdenum (Cairns and Kurtzinger, 1992). The findings are in consonance with the role of molybdenum as a cofactor of aldehyde oxidase, which catalyzes the conversion of abscissic aldehyde to abscissic acid (Seo and Koshiba, 2002). Abscissic acid is responsible for suppressing premature germination through the induction of primary dormancy. High molybdenum content of soybean seeds had earlier been shown to benefit the establishment of seedlings and seed yield of plants grown on low molybdenum soils (Gurley and Giddens, 1969).

CHAPTER 7

BORON

7.1 GENERAL

Boron is a subgroup III metalloid. Boron atoms exist in three valency states and have a strong affinity for oxygen. Unlike other micronutrients, essentiality of boron is confined to vascular plants. Boron (boric acid) in plants has the property to form complexes with hydroxyl radicals of compounds having two closely situated -OH groups in *cis*- configuration. Important among these compounds are the *o*-diphenols and sugars (Loomis and Durst, 1991). Strong boron-diol complexes are formed between boric acid and *cis*-diol furanoid groups of sugars such as apiose and fucose. This property of boron forms the basis for a structural and functional role of boron in plant cell walls and plasma membranes.

The role of boron in bridging the hydroxyl groups accounts for the presence of a large number of naturally occurring boron-containing compounds in plants (Dembitsky et al. 2002, references therein). Biological significance of most of these is, however, not known. Recent researches on the role of boron in plants has led to the establishment of its role in cross-linking of cell wall polysaccharides (O'Neill et al. 2004), which explains

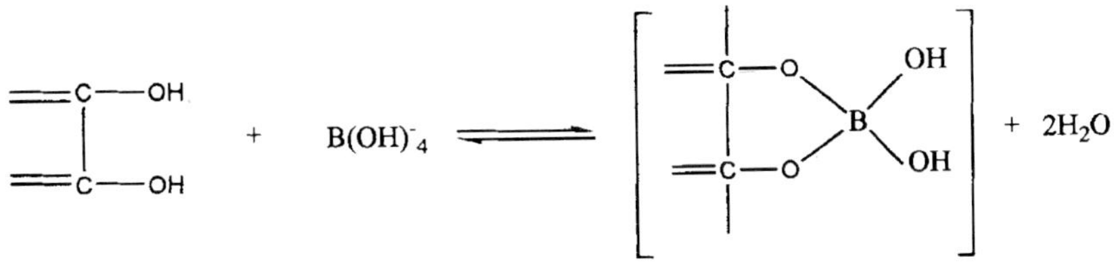

Cis-diol   Borate   Borate easter

Fig 7.1. Complex formation between *cis*-diol furanoid groups of sugars and boron

many physiological responses and symptoms of boron deficiency. Boron deficiency responses of plants are, however, much wider (see reviews by Goldbach, 1997; Blevins and Lukaszewski, 1998, Brown et al. 2002; Lauchli, 2002) than can be explained in terms of its structural role as a constituent of the cell wall.

7.2 UPTAKE, TRANSPORT, DISTRIBUTION

Boron is present in soil solution in several different forms-BO_2^-, $B_4O_7^-$, BO_3, $H_2BO_3^-$ and $B(OH_4)$. The most important form of these is the soluble undissociated boric acid ($B[OH]_3$). At the common cytoplasmic pH (7.5), more than 98% of boron is reported to exist as free boric acid and less than 2% as the borate anion ($B[OH]_4^-$) (Woods, 1996).

Dependence of boron concentration in excised barley roots on the concentration of free boric acid in the nutrient solution led Oertli and Grgurevic (1975) to suggest that plant uptake of boron involves a passive process. Theoretical considerations also supported this contention (Raven, 1980). Studies on boron uptake by sunflower and squash and cultured tobacco cells by Brown and Hu (1994) provided convincing evidence to show that boron uptake involves passive diffusion followed by rapid formation of boron complexes which contributed to a concentration gradient. While it is widely accepted that boron is taken up by a passive mechanism (Power and Woods, 1997; Hu and Brown, 1997; Nable et al. 1997; Dordas and Brown, 2000), results reported by several workers show that boron uptake also involves an active metabolic process (Neales, 1971; Shu et al. 1991; Hu and Brown, 1997). Dannel et al. (1997) and Pfeffer et al. (1997; 1999a) presented evidence to suggest that boron uptake involves both passive and active mechanisms, the predominant mechanism being determined by the availability of boron. When plants are adequately supplied with boron, its uptake is passive, but when boron is in short supply, it is taken up by an active, metabolic process (Pfeffer et al. 1999b). It has been shown that boron deficiency induces an energy dependent, high-affinity transport system which contributes to a boron concentration mechanism in roots (Pfeffer et al. 1999b; Dannel et al. 2000). Studies involving boron transport across vesicles isolated from squash roots suggested that boron uptake is mediated partly through the lipid bilayer and partly through the proteinaceous channels in the plasma membrane (Dordas and Brown, 2000; Dordas et al. 2000). Dordas et al. (2000) also suggested the possible involvement of some major intrinsic proteins in the uptake process. In a recent review, Dannel et al. (2002) have suggested that transport of boric acid across the plasmalemma involves three molecular mechanisms: (i) diffusion across the lipid bilayer; (ii) facilitated permeation through channel-like major intrinsic proteins, that contributes to a passive, constitutive mechanism; and (iii) an energy dependent high-affinity transport system, induced in response to low boron supply (active process). Active uptake of boron is also supported by inhibition of boron uptake by metabolic inhibitors and also by short-term uptake studies, which show that boron uptake at low boron supply follows Michaelis-Menten kinetics characteristic of high-affinity uptake (Pfeffer et al. 2001).

Long-distance transport of boron through the xylem is related to transpiration rates (Raven, 1980). Boron is quite mobile in phloem (Oertli and Richardson, 1970; Oertli, 1993).

Retranslocation of boron to sites that do not lose water rapidly (inflorescence, fruits, etc.) involves phloem (Shelp et al. 1995). Measurement of boron flux in xylem and phloem fluids during fruit development of *Arachis hypogaea* and *Trifolium subteraneum* suggested that boron demand of fruits and storage organs is almost exclusively met through phloem transport. Studies using the stable isotopes ^{10}B and ^{11}B (Marentes et al. 1997) showed that retranslocation of boron to the developing sinks predominantly takes place through phloem and that this involves boron-diol complexes as transport molecules (Hu and Brown, 1996; Hu et al. 1997). Hu and Brown (1996) showed that differences in phloem transport of boron in different species of *Pyrus*, *Malus* and *Prunus* are related to their sorbitol concentration. Species rich in sorbitol are relatively more efficient in phloem transport of boron. Boron-polyol complexes have been isolated and characterized from several plant species (Hu et al. 1997). Phloem sap of celery was shown to contain a boron-mannitol complex, and the peach nectar a boron-sorbitol complex. Close correspondence between boron transport and the boron-polyol complexes in phloem strongly suggests that boron is transported through phloem in the form of boron-polyol complexes. Species differences in the polyols suggest that the specific polyol molecules involved in phloem transport of boron may differ in different species. On the basis of the presence of large quantities of pinitol in soybeans and its responsiveness to boron fertilization, Blevins and Lukaszewski (1998) suggested that in soybeans, phloem transport of boron takes place in the form of boron-pinitol complex. It has been shown that enhanced synthesis of sorbitol in tobacco transgenics facilitates phloem boron transport and enhances tolerance to boron deficiency (Bellaloui et al. 1999; Brown et al. 1999).

Translocation of boron from roots to shoots, as also the intracellular compartmentalization of boron are reported to be influenced by the level of boron supply (Dannel et al. 1998; Pfeffer et al. 1997, 2001; Noguchi et al. 2000; Li et al. 2001). Boron-sufficient plants retain a larger proportion of boron in roots than boron-deficient plants. Short-term deprivation of boron causes more marked decrease in boron concentration in shoots than in roots (Dannel et al. 1998; Li et al. 2001). Boron inefficiency of tomato mutant

T 3238 (Brown and Jones, 1971) and *Arabidopsis* mutant *bor-1-1* (Noguchi et al. 1997, 2000) has been attributed to limitation in transport of boron from roots to shoots under boron deficiency. Pfeffer et al. (2001) have shown that unlike boron-sufficient plants, the roots of boron-deficient plants contain no boron in the free space. Pfeffer et al. (2001) attributed this to possible induction of high-affinity B-uptake mechanism involved in boron transport from free space in to the inside of the cell. While a relatively large proportion of the total cell boron has been shown to be localized to the cell wall as water-insoluble residue (Hu and Brown, 1994; Pfeffer et al. 2001), its compartmentation is reported to be determined by the level of boron supply, the plant organ and the genotype (Dannel et al. 2002).

7.3 ROLE IN PLANTS

7.3.1 Cell Wall Structure

The property of boron to form diester bonds with diol groups of polysaccharides and occurrence of a large proportion of boron in cell walls (Loomis and Durst, 1992) suggested a role of boron in cell walls. Studies with cultured cells (Matoh et al. 1992; Hu and Brown, 1994) in which the possibility of inclusion of extraneous boron in cell wall preparations is minimized, showed that boron is present in the cell walls as an integral component of their polysaccharide complexes and not just bound externally to the cell wall polysaccharides. In cultured tobacco BY-2 cells as much as 92% of cellular boron contained in cell walls is complexed to it as B-polysaccharides (Matoh et al. 1993). Hu and Brown (1994) showed that the cell wall boron in cultured cells of tobacco and squash was closely associated with cell wall pectins and suggested a role of boron as a constituent of cell walls. Hu et al. (1996) analyzed as many as 24 species, differing in boron requirement, for cell wall pectin and inferred that differences in boron requirement of plants are related to differences in the pectin content of cell walls. Significant contributions to role of boron in cell wall structure were made by Matoh and coworkers (Matoh et al. 1993, 1996; Kobayashi et al. 1995, 1996) and O'Neil and coworkers (O'Neill et al. 1996; Pellerin et al. 1996) and Ishii and Matsunaga (1996). Their findings provided evidence to show that the rhamnogalucturonan II component of the cell wall pectic polysaccharides provides a site for boron diester bond formation leading to the formation of boron-rhamnogalacturonan II complex (B-RG II). Boron, as boric acid, was shown to function in covalent bonding of the two chains of monomeric rhamnogalacturonan II to produce dimeric RG II complex which forms an essential constituent of higher plant cell walls (Kobayashi et al. 1996; Matoh et al. 1996; Matoh and Kobayashi, 1998; O'Neill et al. 1996). O'Neill et al. (1996) showed that boron, as boric acid, functions in covalent linking of two chains of rhamnogalacturonan II

(RG II) to produce dimeric RG II complex. The borate ion binds to two apiose residues of two RG II monomers to form a borate diester bond. Fleischer et al. (1999) showed that when cells of *Chenopodium album* L. were cultured on boron-deficient medium for over an year, they contained monomeric rhamnogalacturonan II (m RG-II) but no borate ester cross linked RG-II dimer (d RG-II-B) and relatively large-sized pores in the cell wall. When boric acid was added to the cultures, it was rapidly (within 10 min) bound to the cell walls, d RG-II was produced and cell wall pore size was reduced. The boron-dependent changes were ascribed to formation of boron ester cross-linked pectic network in the primary cell wall.

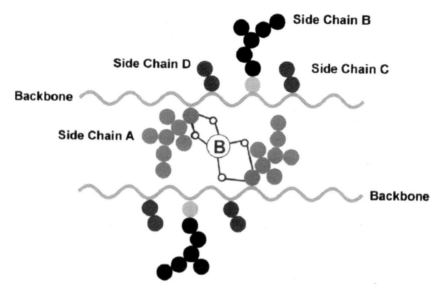

Fig 7.2. Boron cross-linking of two rhamnogalacturonan II (RGII) molecules, each possessing four side chains-A,B,C and D. The aposyl residues in the side chain A of the two molecules become covalently cross-linked by a 1:2 borate diol-ester. (Reprinted from *Annu.Rev.Plant Biol.* Vol.55 ©2004, with permission from Annual Reviews, www.annualreviews.org.)

The borate cross linking of the cell wall RG II (Fig. 7.2) provides the structural organization to the cell walls needed for turgor-driven growth of plant cells (O'Neill et al. 2001). The leaf cell walls of the *Arabidopsis thaliana* mutants *mur 1-1* and *mur 1-2*, that are dwarfed and have small rosette leaves that do not grow normally, contain normal amounts of RG II but only half of it exists as a boron cross-linked dimer. Foliar application of a solution containing boric acid and L-fucose enhances dimerization of RG II, leading to turgor-driven growth of the rosette leaves. Rescue from the L-fucose deficiency defect makes *mur 1-1* mutants resemble the wild type. In a recent study, Ryden et al. (2003) have shown that boron addition to the fructose biosynthesis defective mutant *mur 1* rescues the poor mechanical strength of the hypocotyls and inflorescence stems by contributing fucosyl residues to the pectin network of the cell wall via the RG II- borate complex. Based on these studies, Ryden et al. (2003) suggested a role of RG II- borate complex in providing tensile strength to cell walls.

Kobayashi et al. (1997) showed that, to some extent, germanic acid could substitute for boron in the formation of dimeric RG II, which provided an explanation for the sustained growth of cell lines of tomato in culture medium wherein boric acid was substituted by germanic acid (Loomis and Durst, 1992). Recent studies by Ishii et al. (2002) are not supportive of a role of germanium in cross-linking of RG II monomers. Ishii et al. (2002) examined the effect of boron deprivation on rhamnogalacturonan II in pumpkin (*Cucurbita moschata* Duchesne). The leaf cell walls of boron-deficient plants contained less than one half of the borate cross-linked rhamnogalacturonan II (RG II) dimer found in boron-sufficient plants. Foliar application of [10]B-enriched boric acid increased the proportion of dimeric RG II in the leaf cell walls. Application of germanium failed to produce any such effect, showing that germanium does not substitute for boron in cross-linking of RG II and that boron requirement for cross linking of RG II in the cell walls is absolute.

7.3.2 Membrane Integrity and Function

Boron deficiency effect on ion uptake by plant roots suggested a role of boron in membrane function (Robertson and Loughman, 1974). Robertson and Loughman, (1974) and Pollard et al. (1977) showed reversal of boron deficiency induced change on ion uptake on recovery from boron deficiency. Boron deficiency induced decrease in uptake and efflux of ^{32}P and ^{14}C glucose by sunflower roots and carrot cell cultures is rapidly reversed on resupplying boron and inhibited by vanadate. (Goldbach, 1985). Blaser-Grill et al. (1989) provided direct evidence of boron effect on membrane potential (E_o) of roots and H^+ extrusion by roots of carrot and suspension cultured cells of *Elodea densa* and *Helianthus annuus*. They demonstrated depolarization of cell membranes on transfer of *Elodea densa* leaflets and *Helianthus annuus* root segments from boron sufficient to boron deficient medium and reversal of the effect on restoration of boron supply. Similar observations were made by Ferrol and Donaire (1992) and Raldin et al. (1992). Their findings suggested a role of boron in generating a proton gradient across the plasmalemma. Decrease in proton extrusion by microsomal membranes isolated from the roots of boron-deficient plants was associated with decreased activity of H^+ATPase (Ferrol et al. 1993). In consonance with this, addition of boron to cultured carrot cells increased plasma membrane electron transport and proton secretion (Barr et al. 1993). Using electrophysiological techniques, Schon et al. (1990) demonstrated a direct effect of boron on the membrane potential (E_m) of sunflower root tips. Hyperpolarization of the root tip plasma membrane within 3 minutes of addition of boron to boron-deficient root tips showed that the boron effect on cell membrane permeability is rapid and can account for the changes in ion uptake induced by boron deficiency (observed earlier). Rapidity of the change in membrane function was attributed to the property of boron to form diester bonds with diol groups of membrane polysaccharides (Parr and Loughman, 1983). In a recent article, Brown et al. (2002) have proposed a model depicting boron involvement in formation and functioning of 'membrane rafts' and hypothesized that the role of boron in cellular membranes involve its role in formation, stability and function of the membrane rafts. These domains contain abundance of molecules having cis-diols in appropriate positions (Simons and Ikonen, 1997), providing sites for forming complexes with boron. Boron has also been suggested to be involved in maintenance of membrane potential by reacting with redox enzyme systems (Barr et al. 1993; Lawrence et al. 1995).

While boron effect on membrane function is well evidenced, there is a lack of evidence on the presence of boron as an integral constituent of plasma membrane. Using a spectrofluorometric method for detection of boron, Tanada (1983) reported the occurrence of boron in mungbean protoplasts, but use of a sensitive, microscale method for boron determination failed to show significant amounts of boron in the membranes of cultured tobacco BY-2 cells (Kobayashi et al. 1997).

7.4 DEFICIENCY RESPONSES

7.4.1. Visible Symptoms

As would be evident from the several descriptive names given to symptoms of boron deficiency (Table 10.1), plants show wide variation in boron deficiency symptoms. In general, the major targets of boron deficiency are the terminal growths and the soft conducting tissues (phloem) of the stem. Symptoms are almost always associated with anatomical or histological abnormalities. Generally, symptoms of boron deficiency appear first in terminal shoots in the form of small, deformed leaves, borne on severely condensed branches. The growth of the apical meristems is arrested and the shoot apex turns necrotic, giving rise to multiple axillary branches, which meet the same fate and eventually turn necrotic. This gives the boron-deficient plants a bushy appearance. Stem of boron-deficient plants also turns brittle and shows longitudinal splits of the cortex. There are large differences in sensitivity of plants to boron deficiency. Most cruciferous, leguminous and solanaceous crops show high susceptibility to boron deficiency.

Leaves of cabbage plants subjected to boron deficiency develop dull green water-soaked areas, which turn translucent owing to disintegration of mesophyll tissue. Under severe deficiency, the growing point of the shoot turns necrotic and heads are not formed. Under moderate deficiency, the outer leaves of the head appear thick and brittle. Cauliflower shows somewhat similar symptoms. Under moderate deficiency, the outer leaves appear thick and brittle and leaf petiole and midrib develops small blisters and cracks. Curds of boron-deficient plants develop small water-soaked areas, which later turn brown and necrotic (browning of curd). The interior of the stem also develops water-soaked areas, which eventually turn brown and necrotic. Decay of soft tissues (phloem) produces large cavities, turning the stem hollow (hollow stem). A conspicuous feature of boron-deficient leaves of cabbage and cauliflower is a thick layer of epicuticular wax that gives the leaves a bluish appearance. In lettuce, margins of young leaves become scorched and ragged and the leaves appear 'hooked' and distorted.

In root crops such as turnip, radish and sugarbeet, discolouration, browning and curling of the young terminal leaves (crown) form early symptoms of boron deficiency. As deficiency persists, the basal part of the leaf petiole and the growing point turns dark brown and necrotic and the areas close by decay out (crown rot). The root surface develops wrinkles and cracks, and the softer tissue of the core develops water-soaked areas resulting in tissue disintegration.

In most fruit trees, boron deficiency leads to shortening of internodes and malformation of leaves of young shoots. Under severe deficiency, the terminal shoot may 'die-back' and the terminal buds remain rudimentary. Stem shows splitting of the bark. Fruits have a rough surface, develop cracks and are often malformed.

Unlike the other cereal crops, maize is sensitive to boron deficiency. Severe symptoms of boron deficiency are produced in hybrid maize subjected to boron deficiency in sand culture (Sharma et al. 1979a). Elongation of internodes is severely restricted and young leaves develop white irregular areas that enlarge and coalesce intervenally. Lamina also splits longitudinally. The apical margins of the old leaves become dry, shrivelled, scorched and necrotic. Leaves also give a bronzed appearance and may exude some water-soluble compounds which, on drying, form a crust on the leaf surfaces.

7.4.2. Cell Wall and Membranes

High net efflux of K^+ and of phenolics, sucrose and amino acids in response to boron deficiency indicates increased leakiness of the plasma membranes (Cakmak et al. 1995; Tang and De La Fuente, 1986).

Making use of specialized techniques, several workers have investigated the effect of short-term deprivation of boron on structural changes in cell walls. Yu et al. (2002) made an immunofluorescence microscopic study of the distribution pattern of cell wall pectins in meristematic cells of root apices of maize and wheat plants following boron deprivation. Under boron deficiency, pectins show enhanced accumulation because of restricted internalization. Recently, Yu et al. (2003) investigated the effect of boron deprivation on the cytoskeletal proteins in maize and zucchini root apices. Using immunofluorescence microscopy and western blot, they showed that in maize, short-term deprivation of boron led to increased accumulation of actin and tubulin, associated with alteration in the polymerization pattern of the cytoskeletal assemblies. Rapid cytoskeletal changes in response to boron deprivation suggested a role of boron at translational or post-translational level. Yu et al. (2003) proposed that accumulation of cytoskeletal proteins under boron deficiency is an adaptive response, which contributes to mechanical reinforcement of the cells of root periphery.

Changes in membrane function such as proton extrusion (Blaser-Grill, et al. 1989; Ferrol, 1993) and enhanced efflux of ions and other solutes (Robertson et al. 1974; Tang and De La Fuente, 1986; Pollard et al. 1977; Cakmak et al. 1995) induced in response to boron deficiency indicate boron involvement in maintenance of membrane integrity. Enhanced accumulation of phenolics (Cakmak et al. 1995) and increased activity of phenol-oxidizing enzymes such as polyphenol oxidase (Pfeffer et al. 1998) in boron-deficient tissues may lead to increased production of quniones. The quniones are known to react with oxygen to generate reactive oxygen species (Appel, 1993) and cause peroxidative damage to membrane constituents. Cakmak et al. (1995) suggested a role of boron in preventing the oxidation of phenols by forming boron-phenol complexes. While changes in membrane permeability under boron deficiency are well substantiated, observations made by Pfeffer et al. (1998) and Cara et al. (2002), do not support enhanced oxidation of phenols to be the cause of loss of plasma membrane integrity. Cara et al. (2002) showed that enhanced accumulation of phenols following boron deprivation was associated with a decrease in polyphenol oxidase activity. Functional changes in plasmalemma, such as the activities of membrane-bound ATPase and ferric-chelate reductase were observed only under conditions of severe boron deficiency, subsequent to loss of membrane integrity (Cara et al. 2002).

7.4.3. Photosynthesis, Carbohydrates

Boron deficiency has been shown to decrease photosynthetic O_2 evolution and photosynthetic efficiency of sunflower leaves (Kastori et al. 1995). Boron deficiency induced decrease in the rate of PS II electron transport and quenching of fluorescence emission by chlorophyll molecules (at 655 nm) has been attributed to oxidative damage caused to the thylakoid membranes (El-Shintinawy, 1999).

Boron nutrition of plants affects the metabolism of carbohydrates through its effect on the activities of enzymes catalyzing the interconversion of sugars. The pentose phosphate pathway, which forms an alternative to glycolysis, is particularly sensitive to changes in boron status of plants. This pathway consists of two phases; the first phase involves the oxidative generation of NADPH and the second, synthesis of 5-carbon sugars involving non-oxidative interconversions. In the first reaction of the oxidative phase, catalyzed by glucose 6-phosphate dehydrogenase, glucose 6-phosphate is oxidized to 6-phosphogluconate. The 6-phosphogluconate thus formed is then oxidatively decarboxylated by 6-phosphogluconate dehydrogenase to ribulose 5-phosphate.

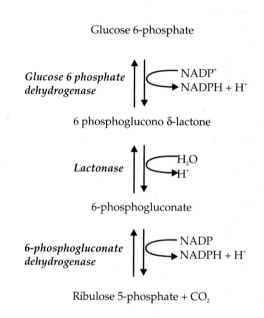

Glucose 6-phosphate

Glucose 6 phosphate dehydrogenase   NADP$^+$ NADPH + H$^+$

6 phosphoglucono δ-lactone

Lactonase   H$_2$O   H$^+$

6-phosphogluconate

6-phosphogluconate dehydrogenase   NADP   NADPH + H$^+$

Ribulose 5-phosphate + CO$_2$

Fig. 7.3. The oxidative pentose phosphate pathway

In boron-deficient plants, the activities of both glucose 6 phosphate dehydrogenase and 6-phosphogluconate dehydrogenase increase (Lee and Aronoff, 1967; Gomez-Rodriguez et al. 1987b). Lee and Aronoff (1967) attributed the increase in 6-phosphogluconate dehydrogenase activity in boron-deficient plants to the property of borate to form complex with 6-phosphogluconate (B-6-phosphogluconate), rendering it inactive to function as substrate for 6-phosphogluconate dehydrogenase. Gomez-Rodriguez et al. (1987b), on the other hand, attributed high activity of 6-phosphogluconate dehydrogenase in boron-deficient plants to associated increase in the activity of glucose 6-phosphate dehydrogenase, causing increased production of its substrate.

Boron-deficient plants show diverse changes in the concentration of carbohydrates. The effect on the different carbohydrate fractions is possibly influenced by the severity of deficiency, the developmental stage of the plant and plant organs. Leaves of boron-deficient plants often show increased accumulation of sugars (Saeed and Woodbridge, 1981). Boron effect on non-structural carbohydrates during the reproductive phase seems to be determined by sink capacity. Poor fertilization and seed set under boron deficiency severely limit sink capacity, causing increased accumulation of photoassimilates in the source leaves, even to levels that inhibit photoassimilation.

7.4.4. Nitrogen Assimilation, Nucleic Acids

There are several reports of inhibition of nitrate reductase activity and enhanced accumulation of nitrate in boron-deficient plants (Bonilla et al. 1980; Kastori and Petrovic, 1989; Ramon et al. 1989; Shen et al. 1993; Camacho-Christobal and Gonzalez-Fontes, 1993). These changes could result from possible involvement of boron in synthesis of the enzyme protein or facilitation of nitrate uptake (Ruiz, 1998a).

Boron deficiency induced changes in the concentration of nucleic acids and enzymes of nucleic acid metabolism suggest its involvement in nucleic acid metabolism (Dugger, 1983). Many workers reported a decrease in nucleic acids, in particular ribonucleic acid, in plants deprived of boron (Johnson and Albert, 1967; Shkolnik, 1984). Johnson and Albert (1967) reported alleviation of boron deficiency symptoms on addition of RNA or individual purines and pyrimidines to nutrient solution. Decrease in RNA content in boron-deficient plants was attributed to inhibition of RNA synthesis. But, high turnover of RNA in boron-deficient plants fed with radioactive precursors of RNA (Cory and Finch, 1967; Chapman and Jackson, 1974) did not substantiate this. Decrease in the RNA content under boron deficiency is possibly a consequence of its enhanced degradation resulting from activation of ribonuclease (Shkolnik, 1984; Dave and Kannan, 1980; Sharma and Abidi, 1986; Chatterjee et al. 1989; Agarwala et al. 1991).

7.4.5. Phosphate Metabolism

Boron deficiency induced changes in pentose phosphate pathway have been described earlier (Section 7.4.2). Several workers have described an increase in the activity of acid phosphatases in boron deficient plants (Hewitt and Tatham, 1960; Hinde and Finch, 1966; Sharma and Abidi, 1986; Agarwala et al. 1991). Restoration of boron supply reverts the enzyme activity to the level of control plants. It is likely that boron is involved in enzymatic regulation of the pool of phosphate esters. Galloway and Dugger (1990) have reported boron-inhibition of phosphoglucomutase.

7.4.6 Phenolics

One of the most common responses to boron deficiency is enhanced accumulation of phenolic compounds. Early work on boron effect on phenolic compounds and possible explanations thereof were reviewed by Dugger (1983) and Pelbaum and Kirkby, (1983). Enhanced accumulation of phenolic compounds in boron-deficient plants could result from activation of pentose phosphate pathway (Lee and Aronoff, 1967; Dugger, 1983; Gomez-Rodriguez et al. 1987b) or increased activity of enzymes involved in the biosynthesis of phenylpropanoids. Boron deprivation of plants increases the activities of polyphenol oxidase (Pfeffer et al. 1998; Camacho-Christobal et al. 2002) and phenylalanine ammonia lyase (Cakmak et al. 1995; Ruiz et al. 1998 b; Camacho-Christobal et al. 2002). Boron deprivation not only induces enhanced accumulation of the existing phenolics but also synthesis of new phenolic compounds (Camacho-Christobal et al. 2002, 2004). In a recent study, Camacho-Christobal et al. (2005) reported increased accumulation of putrescine in tobacco plants deprived of boron. Accumulation of phenolic compounds, particularly caffeic acid and quinones, which are highly reactive, leads to enhanced generation of the superoxide ions (O_2^-), which are known to cause peroxidative damage to cellular membranes. Cakmak et al. (1995) have ascribed the changes in membrane structure and function under boron deficiency to peroxidative damage resulting from enhanced accumulation of the reactive phenolic compounds. The enhanced accumulation of the phenolics has also been suggested to elevate the endogenous levels of auxin (IAA) to toxic limits. Shkolnic et al. (1981) attributed necrosis of the apical growths in boron- deficient plants to increase in auxin concentrations. High build up of phenolics under boron deficiency could inhibit IAA oxidase (Coke and Whitingham, 1968; Rajaratnam and Lowery, 1974; Shkolnic et al. 1981), preventing its oxidative breakdown. However, there is little evidence to show boron involvement in regulation of auxin concentration. Hirsh and Terry (1980) pointed out the differences between ultrastructural changes in roots of sunflower—induced in response to boron deficiency—and auxin application. Endogenous level of auxin in root and shoot tips also show little difference in response to boron deficiency (Hirsh et al. 1982; Fackler et al. 1985).

7.4.7. Secondary Metabolism

Boron deficiency induced changes in plants suggest boron involvement in synthesis of lignin (Lewis, 1980a), polyphenols (Watanabe et al. 1964), flavanoids (Carpena et al. 1984), and alkaloids (Srivastava et al. 1985). Dixit et al. (2002) showed increase in curcumin content of rhizomes of turmeric (*Curcuma longa*) grown under boron deficiency and suggested it to be a consequence of diversion of some photoassimilates to serve as intermediates of curcumin biosynthesis.

Boron involvement in metabolism of phenolics and lignin biosynthesis contributes to strengthening of plant defense against pathogenic infections. Possibly, boron deficiency induced changes in metabolism of plants produce a more favourable environment for the pathogens. Exposure of wheat plants to boron deficiency predisposes them to infection by powdery mildew (Schutte, 1967).

7.4.8. Auxin, Ascorbate Metabolism

There are several early reports of boron affecting IAA concentration in plants, but the results reported are not consistent, possibly because the findings are based on long-term observations that may involve indirect effects. Bohnsack and Albert (1977) examined short-term response to boron deprivation on root tips of squash and described a 20-fold increase in IAA oxidase activity in the root tips within 24 h of withholding boron supply. Boron stimulates the activity of auxin-sensitive NADH oxidase (semidehydroascorbate reductase), which catalyzes the transfer of electrons to ascorbate free radical (Barr and Crane, 1991; Barr et al. 1993). Inhibition of NADH oxidase in membranes of boron-deficient plants could explain the reported decrease in the ascorbate concentration (Lukaszewski and Blevins, 1996; Cakmak and Römheld, 1997). Decrease in ascorbate level in boron-deficient plants could also result from enhancement of ascorbate oxidase activity under conditions of boron deficiency (Brown, 1979). Lukaszewski and Blevins (1996) described cessation of root elongation (apparently an auxin-deficiency response) and a concurrent decrease in the ascorbate concentration in boron-deficient squash roots. Addition of ascorbate to boron-deficient culture solutions counteracted the boron-deficiency induced increase in IAA and promoted root growth. This is in consonance with earlier reports of decrease in ascorbic acid content of boron- deficient plants (Steinberg et al. 1955; Agarwala et al. 1977b) and increase in ascorbate concentration following field application of boron (Chandler and Miller, 1946; Mondy and Munshi, 1993). Mondy and Munshi (1993) reported that boron fertilization of potatoes not only increased ascorbate content of tubers but also prevented them from discoloration during storage. The latter is attributed to accumulation of phenolic compounds. Lukaszewski and Blevins (1996) ascribed the decrease in IAA concentration in boron-deficient squash root meristems to ascorbate induced decrease in *p*-coumaric acid, which functions as monophenol cofactor for IAA oxidase activity. However, recent work with transgenic plants provides little evidence for a role of IAA oxidase in regulation of IAA concentration (Lagrimini, 1991, Bandurski et al. 1995; Normanly et al. 1995).

7.4.9. Nitrogen Fixation

Early work by Brenchley and Thornton (1925) showed decreased nodulation and nitrogen fixation by roots of faba bean grown with a boron-deficient nutrient medium. Requirement of boron for the growth of N_2 fixing cyanobacterium (*Anabaena* PCC 7119) and observed decrease in the glycolipid content of heterocyst envelope following boron deprivation (Mateo et al. 1986) suggested a role of boron in N_2 fixation. It was postulated that the inner glycolipid layer of the hetrocyst envelop offered resistance to diffusion of O_2 into the heterocyst and, by doing so, protected the inactivation of nitrogenase (Layzell and Hunt, 1990; Garcea-Gonzalez, 1988, 1991; Bonilla et al. 1990). Boron deficiency induced changes in the heterocyst membrane, favoured greater diffusion of O_2, causing inactivation of nitrogenase and inhibition of N_2 fixation. Yamagishi and Yamamoto (1994) and Bolanos et al. (1996) found the early stages of nodule development in soybean to be particularly sensitive to boron deficiency. Inadequate supply of boron during early stages of nodule development limited both the development of the root nodules and nitrogen fixation (Yamagishi and Yamamoto, 1994). Bolanos et al. (1996) made a study of the boron effect on *Rhizobium*-legume cell-surface interaction and nodule development in pea. In boron-deficient plants, the number of Rhizobia infecting the host cells and the number of infection threads were reduced and the infection threads developed morphological aberrations. Bolanos et al. (1996) suggested the involvement of boron in *Rhizobium*-legume cell surface interaction favouring nodule development. The cell walls of root nodules of boron-deficient plants showing structural aberrations are reported to lack the covalently bound hydroxyproline/proline rich proteins (Bonilla et al. 1997), which contribute to an O_2 barrier, preventing inactivation of nitrogenase and associated decrease in N_2 fixation.

In a group of legumes (that includes soybean, cowpea), wherein nitrogen fixed in the root nodules is exported to the aerial parts as **ureides** (allantoin, allantoate), boron is involved in the catabolism of ureides via the allantoate amidohydrolase pathway. Deficiency of boron leads to decrease in the activity of allantoate amidohydrolase, which catalyzes the oxidative decarboxylation of allantoate to uredoglycine, with concomitant accumulation of allantoate.

7.4.10. Oxidative stress

Boron has recently been implicated in oxidative stress. Boron deprivation of cultured BY-2 tobacco cells causes over expression of early salicylate-inducible glucosyl transferase (ToGT) and glutathione-*S*-transferase (GST) genes (Kobayashi et al. 2004), which may rescue the system against oxidative stress imposed by redox imbalance resulting from boron-deficiency induced changes in cell wall stucture.

7.4.11. Water Relations

Roth-Bejerano and Itai (1981) showed enhancement of stomatal opening in isolated epidermal strips of *Commelina communis*. Sharma et al. (1984a) showed that boron deprivation of cabbage plants led to a decrease in the stomatal opening associated with increase in diffusive resistance and decrease in transpiration rate (Fig. 7.4). Similar observations were made for cauliflower (Sharma and Sharma, 1987) and mustard (Sharma and Ramachandra, 1990). In cauliflower, the boron deficiency effect on leaf diffusive resistance and transpiration could, to a great extent, be reversed

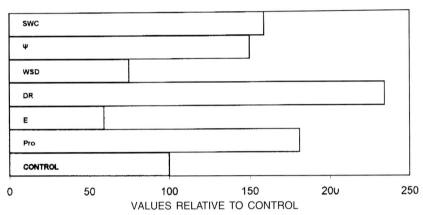

Fig 7.4. Boron deficiency effect (relative values) on specific water content (SWC), water potential (ψ), water saturation deficit (WSD), transpiration (E), diffusive resistance (DR) and proline content (Pro) in cabbage (*Brassica oleracea* L. var. Capitata cv Pride of India) (After Sharma et al. 1984).

on resupply of boron. Observed effects of boron deficiency on tissue hydration (RWC) are not consistent. In early works (Leaf, 1953; Backer et al. 1956), leaves of plants exposed to boron deficiency were reported to have low water content. Opposite results were reported subsequently. Sharma et al. (1984b) and Sharma and Ramachandra, (1990) reported increase in RWC in leaves of boron-deficient plants, but this was associated with decrease in leaf water potential.

Boron-deficient plants also show enhanced accumulation of proline (Sharma et al. 1984b), which is a typical feature of water-stressed plants. The decrease in water potential of boron-deficient plants, inspite of their high tissue hydration, has been interpreted to be caused by its increased partitioning into the bound form, as had been proposed earlier by Backer et al. (1956).

7.4.12. Reproductive Development, Seed Yield

When plants are exposed to moderate deficiency of boron, their reproductive development is inhibited to a greater extent than vegetative development or dry matter production (Gauch and Dugger, 1954; Sharma et al. 1981, Sherrell, 1983; Loomis and Durst, 1992). Relatively high concentration of boron in reproductive parts of flowers such as anthers, ovary and stigma (Gauch and Dugger, 1954; Syworotkin, 1958) and induction of floral abnormalities in boron-deficient plants (Adams et al. 1975; Xu et al. 1993) also point to an involvement of boron in plant reproductive development. Inadequate supply of boron manifests in the form of delayed and restricted flowering, premature bud abscission, pollen sterility, decreased seed set and poor development of seeds. In plants subjected to boron deficiency, both number and size of flowers are severely restricted (Adams et al. 1975; Zhang et al. 1994). Decrease in the number of flowers in boron-deficient plants may be caused by restricted emergence of flower bearing branches and premature bud abscission.

Boron involvement in pollen development and fertility has been central to boron nutrition for a long time. Way back in 1937 (Lohins, 1937) reported atrophy of anthers in boron-deficient plants of several cereal crops. Agarwala et al. (1981) showed that in boron-deficient maize plants, emergence of tassels and anthers is delayed, anthers lack sporogenous tissue and many stamens turn into floral appendages or staminodes. Several other studies substantiate a role of boron in microsporogenesis and male fertility. Boron-deficient plants of oilseed rape show an abnormally developed tapetum (Zhang et al. 1994) and arrest of microsporogenesis beyond the pollen mother cell stage (Xu et al. 1993).

In boron-deficient maize, pollen size and germination percentage is severely reduced (Agarwala et al. 1981; De Wet, 1989). Decrease in pollen germination is observed even before induction of visible symptoms of boron deficiency (Agarwala et al. 1981). Boron-deficient plants of wheat show poor development of anthers and inhibition of floret fertility (Huang et al. 2000). Inhibition of pollen germination due to boron deficiency has also been reported in perennials, e.g. avocado (Smith et al. 1997) and *Picea meyeri* (Wang et al. 2003). Low *in vitro* germination of pollen grains is reflected in poor fertilization (Garg et al. 1979; Agarwala et al. 1981; Huang et al. 2000). Boron deficiency induced cytoskeletal changes in tips of meristematic cells (Baluska et al. 2002; Yu et al. 2002) and reported interaction between cytoskeleton, membranes and cell walls of pollen grains (Li et al. 1997; Franklin-Tong, 1999) suggest possible involvement of boron in cytoskeletal changes preceding pollen germination and pollen tube growth.

Not only does boron play a role in microsporogenesis and pollen germination but it also affects pollen receptivity of the stigma, pollen tube growth through the stylar tract and development of the megagametophyte. The germination of pollen grains on the stigma of boron-deficient *Campsis grandiflora* is reported to be inhibited because of enhanced accumulation of phenolic compounds on the stigmatic surface (Dhakre et al. 1994). In boron-deficient oilseed rape, the stigmatic papilla shows morphological aberrations (Xu et al. 1993) and the rate of pollen tube elongation is retarded (Shen et al. 1994). Based on cross fertilization experiments, Vaughan (1977) had attributed the poor setting of grains in boron-deficient maize plants to non-receptiveness of silks. Robbertse et al. (1990) described a gradient of boron concentration along the style and suggested that this facilitated the growth of pollen tubes. High concentration of boron in stigma and style is also reported to cause inactivation of callose by forming a boron-callose complex at the interface of the pollen tubes and styles (Lewis, 1980b). Boron deficiency is reported to inhibit the development of ovules in cotton (Birnbaum et al. 1974) and oilseed rape (Xu et al. 1993). The boron-deficient plants of oil seed rape are reported to develop abnormal embryo sacs (Xu et al. 1993).

Role of boron in reproductive development of plants has a direct bearing on seed yield (Mozafar, 1993; Rerkasem et al. 1993; Cheng and Rerkasem, 1993; Rawson, 1996a). Low seed yield of wheat in certain areas of warm subtropical Asia is attributed to male sterility caused by reduced transport of boron to the flowering parts, where it is critically required for microsporogenesis and pollen fertility (Huang et al. 1996; Rerkasem, 1996; Rawson, 1996b, Subedi et al. 1998). Boron is also important for post-fertilization development and seed maturation. In sunflower, deprivation of boron supply, even as late as the time of anthesis, produces morphological aberrations in seeds and reduces the seed content of non-reducing sugars, starch and oil (Chatterjee and Nautiyal, 2000). Seeds of low boron sesamum plants show enhanced accumulation of phenolic compounds and decrease in oil content (Sinha et al. 1999). Seeds of low boron black gram plants showed poor germination and lack of vigour (Bell et al. 1989). Foliar application of boron during the reproductive phase of sunflower leads to enhancement in seed yield because of boron involvement in reproductive processes (Asad et al. 2003).

Severe limitation in reproductive development and seed yield of boron-deficient plants could result from several factors restricting the supply of boron to the reproductive organs. Transpiration, and through it the passive delivery of boron to the reproductive organs, may be restricted because of the foliar coverings of the reproductive parts and the inflorescence architecture (Hansen and Breen, 1985) or due to the environmental constraints such as temperature, light, humidity and availability of water during the critical stages of reproductive development (Rawson and Noppakoonwong, 1996; Dell and Huang. 1997). Other major factors that may cause shortfall in boron supply needed for reproductive development are poor translocation of boron from leaves and other mature tissues to the floral parts (Brown et al. 1999) and poor access of the pollen grains and the embryo sacs to the vascular supply (Rawson and Noppakoonwong, 1996; Dell and Huang, 1997; Brown et al. 2002). As the vascular supply to the anther terminates at the tapetum and that to the ovules at the hypostate, the male and the female gametophytes are deprived of access to boron supply directly through the vascular channels.

CHAPTER 8

CHLORINE

8.1 GENERAL

Last to be added to the list of micronutrients (Broyer et al. 1954), chlorine is group VII halogen element. It has only one stable oxidation valency state (Cl⁻) and exists in plants as free anion (Cl⁻), bound to exchange sites or as organic molecules. More than 130 chlorine-containing organic compounds have been reported from plants (Engvild, 1986). Their significance is currently not known, but at least one of these— 4-chloroindole acetic acid— shows high auxin activity (Flowers, 1988). Chlorine is a structural component of the manganese cluster of photosystem II that catalyzes the oxidation of water. Large accumulation of chloride in the vacuoles suggests its involvement in maintenance of turgor and osmoregulation. Chlorine has also been shown to be involved in regulation of enzyme activities and stomatal functioning (Xu et al. 2000).

8.2 UPTAKE, TRANSPORT, TRANSLOCATION

While most soils contain chlorine predominantly as Cl⁻ ions, it is also reported to occur as natural organochlorine compounds (Fleming, 1995; Oberg, 1998). Chlorine is taken up by plant roots either as free anion (Cl⁻) or in association with a monovalent cation. Salts of chlorine are highly soluble and mobile in the rhizosphere. Besides absorption by roots, chlorine is also absorbed by aerial parts of plants, either as an anion (Cl⁻) or as gas (Johnson et al. 1957). Rainwater and environmental pollutants also contribute substantial amounts of Cl⁻ in plants. Different facets of plant uptake and transport of chloride have been reviewed by White and Broadley (2001).

Pitman (1969) suggested chloride uptake across the plasma membranes to involve two transport components, an active component driven by proton pumps (H⁺ ATPases) and a passive component involving facilitated diffusion. The former, coupled to ATP hydrolysis, involves a Cl^-/nH^+ symport. Felle (1994) produced evidence of an electrogenic $Cl^-/2H^+$ symporter in the plasma membranes of *Sinapis alba* root hair cells. The passive component, mediated through anion channels (Tyerman, 1992; Skerrett and Tyerman, 1994), may contribute to major Cl^- influx across the tonoplast. Based on available experimental evidence, White and Broadley (2001) inferred that active transport of Cl^- dominates chloride uptake under low external Cl^- concentrations and passive transport under more saline conditions. In a recent study, using transgenic *Arabidopsis* expressing anion probe, Lorenzen et al. (2004) have presented evidence substantiating that under saline conditions Cl^- influx is passive through channels and that anion transport is coupled with the transport of corresponding cation.

Epstein (1972) proposed that chloride transport across plasmalemma involves multiple-mechanisms with contrasting affinities for Cl. Study of Cl^- fluxes through roots of intact plants and excised root segments shows that the capacity for Cl^- uptake is determined by the plant nutrient status (Pitman, 1969; Lee 1982; Cram 1983).

Root-to-shoot transport of Cl^- takes place through the xylem. Under most conditions, Cl^- loading into the xylem involves the symplastic pathway (Pitman, 1982) but apoplastic transport is also likely (White and Broadley, 2001, references therein). Root-to-shoot transport of Cl^- is influenced by transpiration rate and the rate of shoot growth (Greenway, 1965; Pitman, 1982; Storey, 1995; Moya et al. 1999). Xylem transport of chloride is also influenced by the form of nitrogen supply. Ammonium (NH^+_4) favours and nitrate (NO_3) inhibits transport. Chloride interaction with the form of nitrogen possibly involves its function in maintenance of charge balance.

Chloride is phloem mobile (Lessani and Marschner, 1978). When subjected to salt stress, several species (castor bean, lupin, maize) show relatively high phloem/xylem Cl^- ratio (Flowers, 1988; Jeshke and Pate, 1991; Jeshke et al. 1992; Lohous, 2000). Relatively high concentration of Cl^- in the phloem sap could contribute to phloem loading and unloading of sugars.

8.3 ROLE AND DEFICIENCY RESPONSES

8.3.1 Photosynthesis

Chlorine is a structural constituent of the manganese containing oxygen evolution complex (OEC) of photosystem II. It is postulated to ligand to one or more manganese atoms of the Mn-cluster involved in the donor side reactions of PS II (Renger, 1997; Hoganson and Babcock, 1997). Even before the establishment of Cl as a component of the OEC, chlorine deficiency was shown to decrease O_2 evolution by isolated chloroplasts of higher plants (Bove et al. 1963; Kelley and Izawa, 1978). It has, however, been shown by Terry (1977) that the chlorine effect on photosynthesis in sugarbeet is indirect, and caused by reduction in leaf area resulting from restricted cell division and inhibition of extension growth of cells due to chlorine deficiency. The chlorine-deficient plants of sugarbeet, that show marked decrease in plant growth, show little change in rate of photosynthesis per unit leaf area. Studies on chlorine deficiency effect on stomatal functioning in onion also suggest an indirect effect of chlorine on photosynthetic activity (Schnable, 1980). Proton pumping across the tonoplast causes influx of Cl⁻ simultaneous with K⁺, followed by movement of water into the guard cells, which leads to increase in turgor and opening of stomata, affecting CO_2 exchange.

8.3.2 Osmoregulation

Chlorine has an osmoregulatory role in plants. The level of concentrations at which chlorine functions in osmoregulation is very wide. Chlorine functions as osmotically active solute in roots, affecting plant water relations in excess of 50 mM (Flowers, 1988), which is far in excess of the level at which the other micronutrients play their functional roles in plants. Chlorine needs of other osmoregulatory functions, such as extension growth of cells in root and shoot apices, development of stigma and stomatal functioning, is much less and possibly of the some magnitude as that of the other micronutrients. Chlorine accumulates in relatively high concentrations in the extension zones of root and shoot apices, promoting their turgor-induced extension growth. When deprived of chlorine, maize plants show severe inhibition of root elongation (Hager and Helmle, 1981). The rapid growth of stigma following anthesis in grasses is attributed to extension growth of cells resulting from enhanced cell turgidity following rapid mobilization of K⁺ and Cl⁻ from the neighbouring cells (Heslop-Harrison and Roger 1986). This osmoregulatory role of Cl may have an important bearing on the biology of fertilization.

A study of chloride involvement in stomatal functioning in faba bean leaves by Hanstein and Felle (2002) indicates CO_2 induced enhancement of Cl⁻ efflux from guard cells into the apoplastic fluid. This is attributed to CO_2 induced activation of anion channels of the guard cell plasma membranes. Hanstein and Felle (2002) ascribe the slow nature of the observed response to possible involvement of some CO_2 derived effectors in the activation of the anion channels.

Chlorine has also been shown to be involved in seismonastic movements of *Mimosa pudica* (Fromm and Eschrich, 1988) and circadian rhythm of *Samanea saman* leaflet movements (Kim et al. 1993). Opening of *S. saman* leaflets on exposure to blue light is associated with increased influx of K⁺ and Cl⁻ and consequential changes in turgor of extensor cells. The process is reversed on exposure to red light followed by a dark period. This response, characteristic of a phytochrome-mediated process, suggests that K^+ and Cl^- fluxes across the ion channels of the plasma membranes are light controlled and involve phytochrome action.

8.3.3 Regulation of Enzyme Activities

Even though chlorine is not directly involved in catalysis, it is reported to play a role in regulation of enzyme activities. Rognes (1980) showed chlorine stimulation of asparagine synthetase, which catalyzes the glutamate dependent synthesis of asparagine.

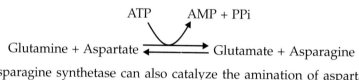

$$\text{ATP} \quad \text{AMP} + \text{PPi}$$

$$\text{Glutamine} + \text{Aspartate} \rightleftharpoons \text{Glutamate} + \text{Asparagine}$$

Asparagine synthetase can also catalyze the amination of aspartate to asparagine. Asparagine functions in storage of nitrogen and its transport from source to sink. It is the major nitrogenous constituent of the phloem sap of several legumes.

8.3.4. Deficiency Symptoms

In controlled culture experiments, avoiding contamination from nutrient solution and the aerial environment (rainfall, pollutants), chlorine deprivation of plants has been shown to retard growth, restrict expansion of young leaves and induce visible symptoms. Chlorine deficiency leads to severe restriction in area of young leaves, possibly due to inhibition in cell division rate. This could also result from a decrease in osmoticum, leading to inhibition in extension growth of cells (Section 8.3.2).

A common feature of chlorine deficiency is intervenal chlorosis or yellowing of young leaves. In most plants, the margins of these leaves invariably turn brown and appear flaccid or wilted (Broyer et al. 1954; Ulrich and Ohki, 1956; Johnson et al. 1957). Wilting of the leaf margins is particularly prominent on hot, sunny days and is possibly related to impairment of osmoregulation. Young leaves of chlorine-deficient plants of red clover curl and shrivel and eventually turn necrotic (Whitehead, 1985). While sugarbeet, lettuce, coconut palm, and kiwi fruit trees show high sensitivity to chlorine-deficiency, seed legumes are relatively tolerant.

PART II

OCCURRENCE, EVALUATION AND AMELIORATION OF DEFICIENCIES

CHAPTER 9

CAUSES AND SPREAD OF DEFICIENCIES

9.1 GENERAL

Plants depend for their micronutrient needs on their availability in the rooting zone. This is a function of their total content, derived from the soil-forming minerals in the parent material, and the several soil chemical and physical properties. The latter determine the dynamics of the equilibrium between the total content of a micronutrient in the soil and its labile pool, from which they are acquired by plant roots. The acquisition of micronutrients from the labile pool is also affected by biological activities in the soil, the physical factors of the environment (temperature, pH, light intensity, etc.) and cultural practices. Genetic attributes and stress disposition of plants also influence the acquisition of micronutrients. Sustained growth and yield of crop plants is a function of the interactive influences of the soil, plant and environmental factors. Any one or more of these factors may adversely affect the availability of micronutrients and reduce it to deficiency concentrations.

Factors influencing the availability of micronutrients to plants have been discussed in general terms in several publications on soils, soil fertility and plant nutrition (Black, 1993; Barber, 1995; Brady and Weil, 1999; Marschner, 1995; Mengel and Kirkby, 2001) and in a more specific manner in publications dealing with one or more micronutrients (Loneragan et al. 1981; Katyal and Randhawa, 1983; Graham et al. 1988; Mortvedt, 1991; Robson, 1993; Gupta, 1993, 1997a; Abadia, 1995; Rengel, 1999). Comprehensive information about the geographic distribution of micronutrient deficiencies the world over is provided in articles by Welch et al. (1991) and White and Zasoski (1999) and in the several publications on particular micronutrients referred above.

Some common factors influencing the availability of the micronutrients are listed in Table 9.1. Factors contributing to deficiency of particular micronutrients have been discussed in subsequent sections.

Table 9.1. Factors affecting the availability of micronutrients to plants

SOIL FACTORS
      Minerals, total content
      Soil reaction (pH), redox
      Cation exchange capacity
      Organic matter
      Nutrient balance
      Moisture
      Aeration
      Soil microorganisms
PLANT FACTORS
      Nutrient uptake efficiency
      Root-shoot transport
      Nutrient accumulation, compartmentalization
      Nutrient utilization, transformation into biogenic molecules,
      Metabolic activity; rate and stage of growth
ENVIRONMENTAL FACTORS
      Light intensity
      Temperature
      Drought
      Flooding; Hypoxia
CULTURAL PRACTICES
      Cultivation practices
      Fertilizer use

9.2 IRON DEFICIENCY

Even though the inherent iron content of most soils is adequate to meet the iron requirement of plants, its deficiency is widespread globally. Most semi arid and arid soils with high base saturation and high oxidation state of its Fe-minerals (ferric hydroxides) are low in available iron. Several factors restrict the availability of iron in soil (Wallace and Lunt, 1960). The more important ones are:

- High pH (calcareous, alkaline soils)
- Low organic matter
- Light soil texture
- Nutrient imbalance (high phosphate, nitrate, heavy metals)
- Water logging (except in lowland rice)
- Low or high temperatures
- Genotypic inefficiency

Iron deficiency is most common on poorly drained calcareous soils containing free carbonates. Even though not lacking in iron content, the calcareous soils, with high bicarbonate content, are low in plant-available iron (Chen and Barak, 1982; Mengel et al. 1984; Mengel, 1994). This is because at high pH, the soluble iron is converted to insoluble hydroxides and rendered unavailable. Increase of pH by one unit decreases the solubility of iron by a factor of 10^{-3}. Grauber and Kosegarten (2002) have shown that an increase in the pH of the leaf apoplast is the primary cause of physiological inactivation of iron in plants grown on calcareous soils. High uptake of nitrate, which contributes to high leaf apoplastic pH (Mengel, 1994; Kosegarten et al. 2001) can be a major cause of Fe-deficiency on calcareous soils. Chlorosis observed in young leaves of plants growing on calcareous soils is commonly known as 'lime-induced chlorosis' (LIC) or simply as 'Iron chlorosis' (IC). Iron deficiency is also observed on high pH sodic soils. Mengel and Geurtzen (1988) have shown a direct relationship between soil alkalinity and development of chlorosis in maize (*Zea mays* L.).

Organic matter in soils binds iron to form organic complexes and serves as a source of plant-available iron. Lack of organic matter in soils may contribute to iron deficiency. Soil conditions favouring biological oxidation of iron ($Fe^{2+} \rightarrow Fe^{3+}$) have a similar effect. Availability of iron is also adversely affected by excess concentrations of the other nutrient elements or even non-essential elements. Manganese (Mn^{4+}) oxidizes Fe^{2+} to Fe^{3+}, reducing its uptake by plants (Paul and Clark, 1996).

Iron deficiency is the most common and widely distributed micronutrient disorder world over. Its intensity is particularly high in the calcareous soils of the semi-arid regions.

9.3 MANGANESE DEFICIENCY

Manganese deficiency is common on soils derived from parent material low in manganese bearing minerals. Other major soil factors that contribute to manganese deficiency are:
- High pH (free carbonate)
- High organic matter
- Impeded drainage
- Overliming of acid soils
- Microbial oxidation

High soil pH strongly inhibits the availability of manganese and constitutes the most common cause of its deficiency on calcareous soils. Increase in soil pH, resulting from overliming, may also reduce manganese availability to deficiency levels. Availability of manganese is also reduced in soils rich in organic matter, particularly so on peats. On low manganese soils, deficiency may also be aggravated by environmental factors.

Soil microorganisms play an important role in influencing the availability of manganese by carrying out redox changes in the rhizosphere. Manganese in soils exists in several oxidation states (Mn^{2+}, Mn^{3+}, Mn^{4+}) but is taken up by plants only as the divalent cation (Mn^{2+}). Microorganisms that oxidize Mn^{2+} to higher oxidation states reduce the availability of manganese and those which reduce the higher oxidation states to Mn^{2+} enhance its availability. Plant root secretions influence the growth of both manganese oxidizing and manganese reducing microorganisms in soils. The nature of these reactions also shows genotypic variations. The root secretions of the manganese efficient genotypes favour the growth of manganese-reducing microorganisms. Allelopathic effects may also influence availability of manganese. Oats form a suitable pre-crop to wheat because its root exudates inhibit the activities of manganese oxidizing microorganisms. This increases the availability of manganese to the subsequent wheat crop, and also provides protection against the take-all disease (Huber and Mc Cay-Buis, 1993). Soybean grown in areas where take-all was severe on the previous wheat crop, contains low concentrations of manganese and may even develop visible symptoms of manganese deficiency (Huber and Mc Cay-Buis, 1993).

Manganese deficiency is common and widely spread on alkaline (high pH) soils of Malta, India, Pakistan, Syria, Italy, Egypt and Lebanon (Sillanpää, 1982). It has also been reported from the agricultural zones of Australia (Donald and Prescott, 1975) and from France and the USA.

9.4 COPPER DEFICIENCY

Copper deficiency in plants is caused by low copper content in the soil parent material and unfavourable influences of soil chemical factors on its availability (Jarvis, 1981). The more important soil factors that contribute to copper deficiency are:
• High organic matter
• High pH
• Coarse texture
Copper released during the weathering of the minerals is adsorbed on the cation exchange sites. In soils rich in organic matter, relatively large proportions of copper released during the weathering are complexed to organic compounds and rendered unavailable for plant uptake (Alloway and Tills, 1984). Availability of copper also decreases with increase in pH. Alkaline soils, rich in organic matter, are particularly susceptible to copper deficiency. Deficiency is common in plants grown on peats. Copper availability to plants grown on light-textured (sandy) soils may be reduced to deficiency levels, particularly where leaching losses are high.

Copper deficiency has been widely reported from Australia, Egypt, the USA and European countries.

9.5 ZINC DEFICIENCY

Except in some coarse-textured acid soils, which may be low in total zinc because of leaching losses, almost invariably, zinc deficiency in plants is caused by reduced availability of zinc. Zinc availability is strongly influenced by soil conditions, environmental factors and cultural practices. Frono et al. (1975b) discussed the importance of soil factors contributing to zinc deficiency in rice, which acquired alarming proportions in south Asian countries during the 1970s. Importance of different factors affecting zinc uptake by plants has been discussed at length by Marschner (1993). The more common soil factor contributing to zinc deficiency are:

- High pH
- Low organic matter
- Coarse soil texture
- High phosphorus fertilization

Solubility of zinc (Zn^{2+}) is highly dependent on pH. Solubility is high at acid pH. With increase in each pH unit, its solubility shows a 100-fold decrease. High pH-induced zinc deficiency is of common occurrence on calcareous soils.

As organic matter in soils provides ligands that bind to Zn^{2+} and subsequently make it available to plants; low organic matter in soils contributes to decrease in its availability. High incidence of zinc deficiency is reported from sandy soils that are poor in organic matter. The peat and muck soils, which contain organic matter in different stages of decomposition, are exceptions to this. High build up of phosphorus in soils, resulting from high application rates of phosphatic fertilizers, may retard zinc uptake by plants and induce its deficiency. Cakmak and Marschner (1986, 1987) reported decreased physiological availability of zinc to cotton plants exposed to excess supply of phosphorus.

Environmental factors such as temperature, moisture and light intensity, which influence zinc acquisition by plants, may contribute to zinc deficiency. Availability of zinc is reduced in cool, wet seasons. Rise in temperature alleviates the deficiency. Nambiar (1976a,b) showed that plants grown on dry soil absorbed more ^{65}Zn than plants grown on wet soils. Marschner and Cakmak (1989) reported an increase in the severity of chlorosis and necrosis in leaves of bean (*Phaseolus vulgaris*) plants exposed to high light intensity. Other likely causes of zinc deficiency are agricultural practices such as use of high analysis fertilizers that carry little zinc as contaminant and intensive cultivation of high-yielding crop cultivars that have relatively high demand on zinc. Preceding crops may secrete allelochemicals that immobilize zinc and restrict zinc availability to crops grown subsequently (Rice, 1984; Saxena et al. 2003).

Zinc deficiency is widespread the world over (Takkar and Walker, 1993, Rashid et al. 1994; Graham and Welch, 1996). It is the single most important constraint in crop production on coarse-textured soils that are low in organic matter and alkaline (with free lime). Zinc deficiency has been reported in a wide variety of crops of the tropical and temperate regions. It is the most common micronutrient disorder limiting crop production in different agro-ecological regions of India (Singh, 2001). Zinc deficiency is reported to be common on calcareous alluvial soils of the middle and lower Yangtse river valley in China (Liu et al. 1990) and on agricultural as well as pasture lands of Australia (Donald and Prescott, 1975; Graham et al. 1992). It is a major problem limiting wheat production in Central Anatolia region of Turkey (Cakmak et al. 1996d).

9.6 MOLYBDENUM DEFICIENCY

The most important soil factors that affect the availability of molybdenum are pH and organic matter. Availability of molybdenum is drastically reduced in acid mineral soils with pH < 5.5. It is particularly so if the soils are rich in iron oxide hydrates, which provides sites for adsorption of molybdate ions (MoO_4^{2-}). In soils rich in organic matter, relatively large proportions of molybdenum may be complexed to soil organic matter, from which it may subsequently be released slowly. Situation is different in peats, wherein a large proportion of molybdenum (MoO_4^{2-}) is reduced and fixed to humic acid, causing a sharp decrease in its availability. This accounts for common occurrence of molybdenum deficiency on low pH peat soils. Different factors affecting plant uptake of molybdenum have been discussed by Gupta (1997b).

The world distribution of molybdenum deficiency closely corresponds to the distribution of acid soils. Deficiency is most common in African countries, particularly Sierra Leone, Zambia, Nigeria and Ghana (Sillanpaa, 1982). Molybdenum deficiency is also widespread in Western Australia, New Zealand, Nepal, Brazil and the calcareous soils along the Yellow river in China.

9.7 BORON DEFICIENCY

Boron availability in soils is largely determined by boron adsorption reactions, the soil properties that affect these reactions and weather conditions (Goldberg, 1997). The main soil factors that contribute to boron deficiency are:
- High pH
- Coarse texture
- Low organic matter
- Low moisture

High pH is the single most important factor limiting boron availability. Except in saline-sodic soils, availability of boron is reduced by increase in soil pH (>6). This effect is particularly marked in calcareous soils, because of enhanced adsorption of borate ions on free carbonates (Goldberg and

Foster, 1991). For the same reason, overliming of acid soils may result in boron deficiency. Boron deficiency is further aggravated by dry weather conditions, which retard the decomposition of organic matter and also prevent the penetration of roots in the soil. High light intensity accentuates boron deficiency by increasing the boron requirement of plants (Tanaka, 1966; Cakmak et al. 1995).

Boron-deficient soils are distributed globally. Deficiency is most prevalent on coarse-textured (sandy) soils of the humid regions, where high leaching losses of boron (e.g. podsols, podzols) further deplete the inherently low boron content of soils to deficiency levels. Soils' analysis reveals boron deficiency to be common in Nepal, Phillipines, India and Thailand (Sillanpaa, 1982). As many as 132 crop species from 80 countries have been reported to respond to boron fertilization (Shorrocks, 1997).

9.8 CHLORINE DEFICIENCY

Chlorides are highly water soluble and leachable. This may reduce Cl^- availability in highly leached mineral soils, particularly where soils have not been previously fertilized with Cl^- containing fertilizers (e.g. KCl).

In general, chlorine deficiency is not a field problem in any part of the world. There are, however, reports of beneficial effects of chlorine containing fertilizers on crops, particularly wheat. This may be attributed to possible involvement of chlorine in reducing the incidence of soil-borne fungal infections (e.g. take-all).

Chapter 10

EVALUATION OF DEFICIENCIES

10.1 GENERAL

Diagnosis of micronutrient deficiencies and evaluation of their severity are a prerequisite for their correction. Early studies on plant nutrition were, therefore, concentrated on development of methods for diagnosis of micronutrient deficiencies. Pioneering contributions in this regard have been made by Wallace (1961), Sprague (1964) and Chapman (1966). The methods for diagnosis of micronutrient deficiencies are essentially based on the following:
a. Foliar symptoms
b. Plant analysis
c. Soil tests
d. Biochemical and spectral changes
e. Crop response to fertilizer amendment

Judicious and sustainable management of deficiency of a micronutrient requires a quantitative assessment of its availability to plants. This is critical for correction of deficiency through fertilizer amendments. The dose of micronutrient amendment has to be chosen in such a way that it raises the availability of the applied micronutrient to give optimal yields but does not build its concentration to a toxic level. It, thus, becomes imperative to not only identify the limiting nutrient but evaluate the severity of its deficiency which determines crop responsiveness to fertilizer application.

10.2 FOLIAR SYMPTOMS

There is a range of concentration within which a micronutrient is required for different cellular and/or physiological functions in plants. If its concentration falls short of the requirement, the cellular functions are impaired and this manifests as retardation of growth and development of visible symptoms. The latter are, to a great extent, characteristic of the limiting micronutrient, and have been suggested to serve as an aid for diagnosing the deficiencies. Some micronutrient deficiency symptoms appeared so characteristic that they were given descriptive names (Table 10.1); some even before they were identified as micronutrient deficiencies. For example, the *wither tip* or *reclamation disease* in wheat grown on reclaimed heath and moorland soils of Netherlands (copper deficiency),

Table 10.1. Common names ascribed to some micronutrient deficiency symptoms

Micronutrient	Plant	Descriptive name
Manganese	Oats (*Avena sativa* L.)	Grey speck
	Sugarbeet (*Beta vulgaris* L. ssp. *Vulgaris*)	Speckled yellows
	Pea (*Pisum sativum* L.)	Marsh spot
	Tung (*Aleuritis* sp.)	Frenching
	Sugarcane (*Saccharum* spp.)	Pahala blight.
Copper	Citrus (*Citrus* spp), deciduous fruit trees	Exanthema,
	Apple (*Malus sylvestris* Mill.)	Summer die-back
	Wheat (*Triticum aestivum* L.)	Wither tip, white tip, reclamation disease
Zinc	Citrus (*Citrus* spp.) fruit trees	Mottle leaf, frenching, little leaf
	Apple (*Malus sylvestris* Mill)	Little leaf
	Pecan (*Carya pecan*)	Rosette
	Cocoa (*Theobroma cacao*)	Sickle leaf, narrow leaf
	Tung (*Aleuritis fordii*)	Bronzing
	Maize (*Zea mays* L.)	White bud
	Rice (*Oryza sativa*)	Khaira
Molybdenum	Cauliflower (*Brassica oleracea* L. var. botrytis)	Whiptail
	Beans (*Phaseolus* spp.)	Scald
	Citrus (*Citrus* spp.)	Yellow spot
Boron	Apple (*Malus sylvestris* Mill.)	Internal cork, corky pith
	Apricots (*Prunus armenica* L.)	Brown spot
	Rasberry	Die-back
	Citrus (*Citrus* spp.)	Hard fruit
	Celery (*Apium graveoleus* L.)	Cracked stem
	Hops (*Humulus lupulus*)	Ivy leaves
	Lucerne, Alfalfa (*Medicago sativa* L.)	Yellows
	Sugarbeet (*Beeta vulgaris* L. ssp. *vulgaris*)	Heart rot, Crown rot, Dry rot
	Cauliflower (*Brassica oleracea* L. var. botrytis)	Browning of curd, Hollow stem
	Tobacco (*Nicotiana tabacum* L.)	Top sickness
	Turnip (*Brassica rapa* var. Rapa L.)	Canker, Internal black spot
	Swede, Rutabaga (*Brassica napus* L. napobrassica group)	Brown heart, Rahn

Pahala blight of sugarcane grown in the Pahala region of Hawaii (manganese deficiency), *Whip tail* of cauliflower grown on acid soils of England (molybdenum deficiency), and the *Khaira disease* of rice grown on sub-montane soils of India (zinc deficiency).

Symptoms of many micronutrient deficiencies have been experimentally produced under greenhouse conditions, often using sand and solution culture techniques (Hewitt, 1966). Illustrated descriptions of many micronutrient deficiency symptoms of plants of economic importance have been provided by Wallace (1961), Sprague (1964), Hewitt (1963), Scaife and Turner (1983), Agarwala and Sharma (1979), Bould et al. (1983), Winsor and Adams (1987), Bergmann (1992), Bennett (1993) and Sharma (1996). Some common plants reported to be sensitive to micronutrient deficiencies are listed in Table 10.2. Typical symptoms of Fe, Mn, Cu, Zn, Mo and B deficiencies in some agricultural and horticultural crops produced in sand culture by the Plant Nutrition group of the Lucknow University are shown in Plates 1 to 6.

While visible symptoms provide an easy-to-follow method for diagnosis of deficiencies, one drawback of this method is that by the time the symptoms become conspicuous enough, the plant reaches a stage at which the deficiency induced structural and functional damage becomes irreversible. In such a situation, the symptoms are useful only in forewarning of recurrence of the disorder in the successive crop(s). Another difficulty with diagnostic use of visible symptoms is that simultaneous occurrence of another deficiency or toxicity may make them atypical.

10.3 PLANT ANALYSIS

The principles and methods for the evaluation of plant nutrient status on the basis of tissue analysis, the interpretation of plant analysis data and possible flaws that may hamper the right decision on the correction of the deficiencies through fertilizer practices, have been addressed exhaustively in several publications (Chapman, 1966; Bould et al. 1983; Reuter and Robinson, 1986; Munsen and Nelson 1990; Bergmann 1992; Jones, 1999). Comprehensive compilations have also been made of critical concentrations of micronutrients in different crop species (Melstead et al. 1969: Reuter and Robinson 1986; Bergmann, 1992). The tissue concentration of micronutrients in plants provide useful information for evaluation of field-grown crops in terms of sufficiency or deficiency. The use of tissue concentration of mineral nutrients in plants, for the quantification of nutrient status was first proposed by Goodall and Gregory (1947) and subsequently developed by Ulrich (1952). Ulrich and Hills (1967) described the method for working out critical nutrient concentrations, on the basis of relationship between the nutrient concentration and the relative yield of plants.

Table 10.2. Common visible symptoms and plants sensitve to micronutrient deficiencies

Micro-nutrient	Common prominent visible symptoms	Deficiency sensitive plants	
Fe	Chlorosis, first appearing in intervenal areas of young leaves. Severe/prolonged deficiency leads to ivory chlorosis or bleaching, followed by brown necrosis of the lamina	*Cereals:*	Sorghum, maize, rice (seedlings)
		Legumes:	Groundnut, beans
		Oilseeds:	Sorghum
		Vegetables:	Radish, knolkhol
		Fruits trees:	Soft fruits, Citrus Grapevine
		Others:	Sugarcane (ratoons)
Mn	Chlorosis and necrosis of sub terminal leaves	*Cereals:*	Wheat, oats
		Legumes:	Peas, beans, soybean
		Vegetables:	Beets, onion
		Fruits trees:	Citrus fruits, apple, cherry
		Others:	Sugarbeer, sugarcane tobacco
Cu	Chlorosis (or bleaching) of young leaves; Young leaves of cereals fail to unroll (needle shaped); Withering and coiling of leaf tips; Wilted appearance of leaves in broad leaved plants	*Cereals:*	Barely, wheat, oats
		Legumes:	Lupins
		Oil seeds:	Sunflower, flax
		Vegetables:	Carrots, lettuce, onion, spinach, aubergine
		Fruits trees:	Citrus fruits
		Others:	Alfalfa
Zn	Condensation of internodes; Young leaves are small and closely packed (rosetted). Subterminal leaves chlorotic, often develop brown or purple coloration. Salt exudations on leaf surface. Premature abscission of leaves, floral buds and young fruits.	*Cereals:*	Rice, wheat, maize
		Legumes:	Beans
		Oil seeds:	Rapeseed mustard, soybean, flax
		Vegetables:	Cucurbits, tomato
		Fruits trees:	Citrus fruits, grapevine
		Others:	Hops
Mo	Yellowing (pale yellow) of leaves; intervenal chlorosis followed by browning and necrosis. Scorching of leaf margins (mature leaves). Cruciferous vegetables show cupping of young leaves, severe loss of lamina in middle leaves (whiptail).	*Cereals:*	Maize
		Oilseeds:	Sunflower, rapeseed mustard
		Vegetables:	Cauliflower, cabbage
B	Cessation and abnormalities of apical shoot growth. Cracks on stem, decay of soft tissues (phloem necrosis). Premature abscission of leaves, flower buds and fruits. Hard fruits	*Cereals:*	Maize
		Legumes:	Alfalfa, beans
		Oilseeds:	Sunflower, rapeseed mustard
		Vegetables:	Brassicaceae (cabbage, cauliflower, radish), beets, spinach, tomato, aubergine, lettuce)
		Fruits trees:	Citrus, apple, pear
		Others:	Alfalfa

Controlled sand and solution culture methods (Hewitt, 1966) came handy in supplying the micronutrients at varying levels through defined changes in nutrient solution. However, the conventional nutrient solutions have a drawback in that these solutions include some nutrients, including the micronutrients, at levels far in excess of their concentrations normally found in soil solution. Several variations and improvements have been made to make the solution culture method more realistic by maintaining, throughout the growth period, the concentrations of the nutrient elements in the nutrient culture close to that encountered in the soil solution (Asher and Edwards, 1983; Parker and Norvell, 1999). The flowing solution technique developed by Edwards and Asher (1974) offers major advantages over the use of the static solution culture methods by providing the control of root temperature, pH and ion concentrations. But in spite of these added advantage; the flowing solution culture method has not found much use in working out the critical concentration limits of micronutrients because of practical difficulties in manipulating the supply of the micronutrients to desired levels, particularly in the deficiency range. The limitation of supplying nutrients at a relatively high concentration in the culture solution for maintaining their adequacy over a long periods of time may be largely overcome by the use of the buffered nutrient solutions (Parker and Norvell, 1999). The difficulty in case of the cationic micronutrients (Fe, Mn, Cu, Zn) has been overcome by the use of buffered solutions that contain synthetic metal chelators, such as DTPA, to 'buffer' the excess free cation activities in the nanomolar range. The formation constant of the metal chelate complex, the excess of the chelator in the nutrient solution, and the composition of the nutrient solution control the activity of the micronutrients in the nutrient solution to desired levels. Chelated buffered nutrient solutions have been developed for the cationic micronutrients—Fe, Mn, Cu, Zn (Chaney et al. 1989; Bell et al. 1991; Parker 1993, 1997; Welch and Norvell, 1993; Yang et al. 1994). Asad et al. (1997) have developed a buffered nutrient solution for boron, wherein a boron specific resin is used for complexing with boric acid. Mixing appropriate quantities of boron-loaded and boron-free resin can supply boron at different levels. Advances in solution culture techniques, including micronutrient-buffered solution, have been reviewed by Parker and Norvell (1999). The buffered micronutrient solutions are currently preferred over the conventional culture solution for screening plant genotypes for tolerance to micronutrient deficiencies and toxicities.

The method of plant analysis for assessing nutrient status in terms of deficiency, sufficiency and toxicity is based on the well-established relationships between nutrient supply and uptake and between nutrient concentration and growth or dry matter production of plants. Within limits, there is a linear and/or exponential relationship between nutrient supply and nutrient accumulation in plants. A plot of nutrient concentration versus plant yield (dry matter production) shows two points of inflection. First, yield increases with nutrient concentration and then it slows down to form a plateau. Second, yield declines with further increase in nutrient concentration. The former corresponds to critical deficiency limit and the latter to critical toxicity limit. These critical limits are best worked out by growing plants under controlled culture conditions in greenhouses at known levels of micronutrients, ensuring that no other nutrient becomes limiting. Plants are grown over a wide range of nutrient supply, increasing several folds starting from a minimum, sustaining early vegetative growth, to excess. At well-identified stages of active growth, samples are drawn for determining biomass and tissue concentration of nutrients in selected plant parts, generally the leaves that develop the first symptoms of the deficiency. The biomass yield, relative to the maximum obtained under the experimental conditions (relative yield) is plotted against the nutrient concentration in plants grown with varying levels of nutrient supply. Generally, increase in the nutrient supply is associated with increase in leaf tissue concentration and biomass yield until the yield reaches a maximum. Increase in nutrient supply beyond this leads to increment in tissue concentration of the nutrient but not the plant yield. Beyond a certain limit, increase in nutrient supply causes increase in nutrient concentration concomitant with yield decrement.

The nutrient concentration in plants corresponding to optimal biomass yield $\pm 10\%$ is taken as the nutrient sufficiency range (NSR). In this range, yield response to change in nutrient supply/concentration is limited to <10%. Nutrient concentrations corresponding to 90% of the optimal yield in the sub-and supra-optimal ranges are taken as the critical concentration for deficiency (CCD) and toxicity (CCT), respectively. Extent of the decrease in the tissue concentration in the sub-optimal range and increase in the concentration in the supra-optimal range denotes the severity of deficiency

Table 10.3. Dry matter yield and leaf tissue concentration of manganese in maize (*Zea mays* L.) grown with varying levels of manganese supply[1]

Mn supply (μ moles L^{-1})	Dry weight yield (g plant^{-1})	Relative yield (% maximum)	Mn-concentration (μg g^{-1}dry weight)
0.2	35.5	52	6
1.0	48.6	71	10
2.0	55.4	81	17
10	68.4	100	68
20	64.3	94	124
100	54.9	80	170

[1] 35 DAS, top 3 leaves

Fig. 10.1. Growth response of maize to varying levels of manganese supply: Arranged left to right are pots supplied with 0.2, 1, 2, 10, 20 and 100 μM Mn L^{-1}

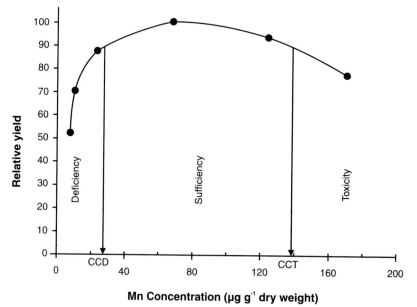

Fig. 10.2. Plot of relative yield of maize (*Zea mays* L.) plants grown at varying levels of manganese supply against tissue concentrations in young (top three) leaves, CCD and CCT denote critical concentration for deficiency and toxicity, respectively.

or toxicity. Plant tissue analysis, thus, helps to quantify the extent of deficiency or toxicity and predicted response to fertilizer amendment.

10.3.1 Iron Deficiency

Plant analysis forms a poor basis for evaluation and management of iron deficiency. Plants with about the same or even higher concentration of iron as in apparently healthy plants may exhibit chlorosis, a feature referred to as 'Fe-chlorosis paradox' (Romheld, 2000). There may be more than one reason for this discrepancy. The field-drawn samples may contain substantial quantities of dust adhering to the leaf surfaces and this may, if not removed prior to determination of tissue iron, contribute varying amounts of iron to the leaf tissue concentration. (Jones, 1992). Removal of surface contamination of dust has been a subject of several investigations. Washing of leaf material with water is of little advantage. A quick wash with a dilute solution of a detergent (0.1%) along with or followed by dilute hydrochloric acid (0.1 m HCl) may prove effective in removal of surface contamination of iron. It has, however, been pointed out that washing of samples may remove a part of endogenous iron, manganese and zinc (Moraghan, 1991).

Even in plants grown under controlled conditions, where dust is not a major problem, iron concentration of plants may show a poor correlation with the level of iron supply (Agarwala and Sharma, 1961). Notwithstanding this, several workers have reported critical concentration values for iron deficiency. These values range from 50 to 150 μg Fe g^{-1} dry weight. In general, critical limits are not of much use in categorizing plants in terms of deficiency and sufficiency or responsiveness to iron amendments. Smith et al. (1984) made a comparative study of iron requirement of some C_3 and C_4 plants and found that even though the C_4 plants have a higher requirement of iron than the C_3 plants, the critical deficiency concentration of iron in the two types of plants is about the some (~ 72 μg g^{-1} dry weight). Tissue concentration of iron in plants receiving iron at different levels of supply may show dissimilar diurnal variations (Sharma and Mehrotra, 1998), which may limit the value of iron concentration as a dependable measure of iron status of plants. Differences in within- the-plant distribution and compartmentalization of iron in normal and iron-deficient plants also limits the suitability of tissue concentration of iron as a measure of iron status of plants.

Over 70 years ago, Oserkowsky (1933) studied the relationship between chlorophyll content and iron concentration in green and chlorotic pear leaves and suggested that the chlorotic leaves accumulate a large proportion of iron that is physiologically 'inactive'. Ever since then, efforts have been on for developing methods for determining what Somers and Shive (1942) termed 'active iron'. These methods involve extraction of iron from leaf tissues in dilute acids or iron–chelates (Katyal and Sharma, 1980; Mehrotra et al. 1985) and finding one which correlates best with chlorophyll concentration. Some workers suggested quantification of Fe(II) iron in plant extracts as a measure of active iron (Katyal and Sharma, 1980; Olsen et al. 1982), but this could be misleading because, during extraction, varying amounts of Fe(III) may be reduced to Fe(II), both photochemically (Krizek et al. 1982) and chemically (Mehrotra and Gupta, 1990). Mohammad et al. (1998) described nitric acid and o-phenanthroline extractable iron in citrus lemon leaves as a measure of active iron that could be used for diagnosis of iron chlorosis. The presently known methods for quantification of active iron have, however, not found to be of much use because of limited information on what constitutes this fraction in terms of the oxidation state of iron and its compartmentalization.

10.3.2 Manganese Deficiency

Manganese concentration of plants forms a suitable basis for evaluation of manganese deficiency and prediction of plant responsiveness to manganese fertilization. Compared to other micronutrients, the range of critical deficiency concentration of manganese in plants is narrow. For most crops, the critical deficiency concentrations range between 10 and 20 μg Mn g^{-1} dry weight.

10.3.3 Copper Deficiency

Copper concentration of plants is lower than for the other cationic micronutrients and difference between its deficiency and sufficiency values is narrow. The critical concentration for copper deficiency is within 1 to 5 μg g^{-1} dry weight (Robson and Reuter, 1981). The copper concentration of young leaves is a more sensitive indicator of copper status than the old leaves (Hill et al. 1979; Loneregan et al. 1981). The critical deficiency limits of copper may also be influenced by the level of nitrogen supply (Thiel and Fink, 1973).

10.3.4 Zinc Deficiency

Zinc concentration in leaves has been widely used as a measure of zinc-nutrient status of plants. Several workers have worked out critical zinc concentrations in a wide variety of crops. In most plants, the critical deficiency concentration of zinc ranges between 15 and 20 μg g^{-1} dry weight. (Robson and Reuter, 1981, Sharma, 1996). However, reports suggest poor correlation between the total concentration of zinc and its physiological availability (Leece, 1978; Ghoneim and Bussler, 1980; Gibson and Leece, 1981; Cakmak and Marschner, 1987, 1993). Rahimi and Shropp, 1984, showed that in several plants of different taxa, water-soluble zinc in leaves served a better indicator of zinc nutrient status than its total concentration.

Similar observations were made by Cakmak and Marschner (1987), particularly in the case of phosphorus-induced zinc deficiency. Limitation in use of total zinc concentration as a measure of functionaly available zinc may be caused by binding of free ionic zinc to cell walls (Youngdahl et al. 1977), an organic ligand or change in apoplast-symplast distribution of zinc.

10.3.5 Molybdenum Deficiency

Deficiency values and critical concentrations for molybdenum deficiency for a wide variety of crop plants have been presented by Gupta (1997c) and Johnson et al. (1997). Generally, the deficiency values range between 0.01 and 0.5 μg Mo g^{-1} dry weight. The critical concentration for molybdenum deficiency shows wide variations. Apart from genotypic differences in molybdenum efficiency, these differences may be caused by differences in the functional requirement of molybdenum. For example, plants that are supplied nitrogen as nitrate require molybdenum for its reduction, which is not the case when nitrogen is supplied in the reduced form (NH_4). Likewise, nodulating legumes require more molybdenum for N_2 fixation than the non-nodulating plants. In nodulating legumes, accumulation of relatively large proportion of molybdenum in root nodules may also affect its concentration in top parts. Reported differences in the deficiency limits of molybdenum for some plants can possibly be attributed to differences in the stage of growth or plant part sampled, and even to the different methods used for quantification of molybdenum (Gupta, 1997c). Presently, there is lack of an accurate, easy-to-follow method for measurement of molybdenum concentration at levels, which may be critical for deficiency.

10.3.6 Boron Deficiency

In keeping with large differences in boron requirement of plants, the critical concentrations for boron deficiency also show a wide range. In graminaceous plants, the critical deficiency values generally range from 5 to 10 μg B g^{-1} dry weight. In the gum-bearing plants such as poppy, the critical deficiency values of boron may be higher (Bergmann, 1992). Kirk and Loneragan (1988) suggested that the relationship between tissue concentration of boron and rate of elongation of the youngest leaf may serve a better basis for working out the critical limits for boron deficiency than the dry matter yield of plants.

10.3.7 Chlorine Deficiency

Chloride content of plants usually varies between 2 and 20 mg g^{-1} dry weight. This is many times higher than for any other micronutrient. Tissue concentration of chlorine at which deficiency symptoms are observed varies between 0.1 and 5 mg g^{-1} dry weight (Xu et al. 2000). Such low concentrations of chlorine can be supplied through rainfall. Since tissue concentration of chlorine in field-grown crops is far in excess of this, deficiency concentrations of chlorine have little relevance.

10.3.8 Tissue Tests

Tissue testing is a rapid method for determining tissue concentration of micronutrients in a field or in laboratories. It involves a semi-quantitative test of plant sap for one or two nutrients. Fresh tissue samples drawn from a selected plant part are put to a semi-quantitative test for a nutrient using rapid colorimetric methods. Such tests have a limited value for evaluating micronutrient status of plants, possibly excepting iron (Chaney, 1984). Studies described under Part I show a relation between accumulation of the micronutrients in the seeds and their concentration in the vegetative parts (leaves) of the mother plants. This relationship forms the basis of the evaluation of plant micronutrient status on the basis of seed analysis. Rashid and Fox (1992) have found zinc analyses of seeds as a suitable method for evaluation of zinc requirement of grain crops. Information about the seed content of micronutrients is of much use in predicting plant performance during the seedling growth, particularly on marginal soils, but its use as a measure of plant nutrient status for diagnostic purpose has limited value.

10.4 SOIL TESTS

Plants acquire micronutrients out of their labile pool in the rhizosphere. The soluble fraction of this pool forms the plant-available micronutrients and can, in different proportions, be extracted by mineral acids, neutral salts, reducing agents or chelates. In order to be used as a measure of availability to plants, the concentration of micronutrients in the extractant should represent a definite proportion of the labile pool, should be large enough to be measurable accurately and reproducibly, and correlate well with the plant nutrient uptake and yield response. In keeping with these considerations, detailed procedures for the quantification of available micronutrients in different soil types and for different crop species have been worked out and described in detail (Jones, 1988; Westermann, 1990; Sims and Johnson, 1991; Bigham, 1996).

Many extractants have been tested to provide an estimation of the available micronutrients (Cox and Kemprath, 1972; Walsh and Beaton, 1972). Use of common extractants, for simultaneous extraction and determination of more than one micronutrient (Lindsay and Norvell, 1978; Soltanpour and Schwab, 1977; Mehlich, 1984; Soltanpour, 1985; Rodriguez et al. 1999) has made a significant advance in the use of soil tests for delineating micronutrient deficient ares. The critical points relating to the extraction procedures and suitability of different soil tests have been discussed at length for the cationic micronutrients by Martens and Lindsay (1990) and Liang and Karamanos (1993), for molybdenum by Johnson and Fixen (1990) and for boron by Gupta (1993). It is important to adhere to the details of the extraction procedures because variations in these may influence the solublization of the labile pool of the soil micronutrients and thus the critical values of the nutrients based on these procedures. The relationship between a chemically extractable nutrient in the soil and its nutrient accumulation in plants may, however, be influenced by several soil and plant factors (Barber, 1995) and this may result in differences in the critical values reported by different workers. There is, however, little doubt that soil tests, using appropriate chemical extractants form a profitable and sustainable method for management of nutrient deficiencies (Soltanpour and Delgado, 2002).

10.4.1 Iron

The commonly used extractants for iron are given in Table 10.4.

Table 10.4.   Common extractants for determining iron availability in soils

• $\underline{N}$ Ammonium acetate, pH 4.8
• 0.5 $\underline{M}$ Ammonium acetate + 0.5 $\underline{M}$ Acetic acid + 0.02 $\underline{N}$ Na$_2$ EDTA, pH 4.65 (AAAc-EDTA)
• 0.005 $\underline{N}$ DTPA + 0.1 $\underline{M}$ Triethanolamine + 0.01 $\underline{M}$ Calcium chloride, pH 7.3 (DTPA-TEA)
• Ammonium bicarbonate + DTPA (AB-DTPA)

Prior to development of common extractants for simultaneous extraction of iron and other micronutrients (Soltanpour and Schwab, 1977; Lindsay and Norvell, 1978; Havlin and Soltanpour, 1981), wide use was made of $\underline{N}$ Ammonium acetate (pH 4.8) extraction, with 2 μg Fe g^{-1} soil as the critical limit for delineation of iron deficient soils. The advantage of simultaneous extraction and determination of all the cationic micronutrients and the suitability of the method for calcareous soils, where iron deficiency is most prevalent, have made DTPA-TEA and AB-DTPA as the methods of choice for determining iron availability. The critical limits for deficiency of iron, using DTPA-TEA extraction range from 2.5 to 4.5 μg g^{-1} soil. Soils with >4.5 μg Fe g^{-1} soil are not expected to show iron deficiency (Lindsay and Norvell, 1978). The deficiency limit for NH_4HCO_3-EDTA extractable iron is about the same – 4.8 μg Fe g^{-1} soil (Havlin and Soltanpour, 1981).

10.4.2 Manganese

Several chemical extractants have been used to assess manganese availability in soils (Table 10.5). Except in acid soils, water-extractable manganese concentration is too low to allow accurate determination. Extraction with ammonium acetate (Browman et al. 1969) is fairly well correlated with plant manganese content. Extraction with hydroquinone in ammonium acetate includes easily reducible manganese and is particularly suited to organic soils. Phosphoric acid and ammonium dihydrogen phosphate extractions are suited for mineral soils.

Table 10.5. Common extractants for determining manganese availability in soils

---

- Water
- $\underline{N}$ Ammonium acetate, pH 7.0
- 0.5% Hydroquinone + 1.0 $\underline{M}$ Ammonium acetate, pH 7.0
- 0.033 $\underline{M}$ Phosphoric acid
- 1.0 $\underline{M}$ or 1.5 $\underline{M}$ Ammonium dihydrogen phosphate
- 0.05 $\underline{N}$ Hydrochloric acid + 0.025 $\underline{N}$ Sulphuric acid (Double acid)
- 0.005 $\underline{N}$ DTPA + 0.1 $\underline{M}$ TEA + 0.01 $\underline{M}$ $CaCL_2$ pH 7.3
- 0.05 $\underline{M}$ EDTA
- NH_4HCO_3-DTPA

Ammonium acetate (pH 7.0) and DTPA-TEA extraction methods are used widely. In the absence of a more extensive study, Lindsay and Norvell (1978) had suggested 1 μg Mn g^{-1} soil as the critical limit for DTPA-TEA extractable manganese but a global study by Sillanpaa (1982) suggested 2 to 3 μg Mn g^{-1} soil to be more appropriate.

10.4.3 Copper

Of the different methods developed for determining the available copper in soils (Table 10.6), the most common and widely used is the DTPA-TEA extraction. The critical limit with this extractant is 0.2 μg Cu g^{-1} soil. Lower values are associated with copper deficiency (Viets and Lindsay, 1973; Lindsay and Norvell, 1978).

Table 10.6. Common extractants for assessing copper availability in soils.

- 0.05 $\underline{M}$ HCl
- 0.05 $\underline{N}$ HCl + 0.025 $\underline{N}$ H_2SO_4 (Double acid)
- $\underline{N}$ Ammonium acetate, pH 4.8
- Ammonium citrate + EDTA, pH 8.5
- 0.05 $\underline{M}$ EDTA
- DTPA-TEA
- NH_4HCO_3-DTPA

10.4.4 Zinc

Several chemical extractants have been developed to determine plant available zinc in soils (Table 10.7). Some of the extractants are specific to zinc, others (Double acid, DTPA, AB-DTPA, AAAc-EDTA) simultaneously extract other cationic micronutrient(s).

Extraction with mineral acids may include part of the non-labile pool. The method is not suited for alkaline soils with free carbonates, where zinc deficiency is most prevalent. Extraction with chelating agents (EDTA, DTPA) offer advantage in the sense that their pH can be buffered close to that of the native soil. This provides better assessment of zinc availability under the prevailing conditions. Extraction with chelating agents is particularly suited to calcareous soils.

Table 10.7. Common chemical extractants for assessing zinc availability in soils

- 0.1 $\underline{N}$ HCl
- 0.05 $\underline{N}$ HCl + 0.025 N H_2SO_4 (Double acid)
- 0.5M Ammonium acetate + 0.5 $\underline{M}$ acetic acid + 0.02 $\underline{M}$ Na$_2$ EDTA, pH 4.65 (AAAc-EDTA)
- 0.02 $\underline{N}$ Magnesium chloride
- 0.01 $\underline{M}$ EDTA + $\underline{M}$ Ammonium carbonate, pH 8.6 0.7–1.4
- 0.05 $\underline{M}$ EDTA, pH 7–9
- 0.01% Dithizone in CCL_4 + $\underline{N}$ NH$_4$OAC, pH 7.0
- 0.005 $\underline{N}$ DTPA + 0.1 $\underline{M}$ Triethanolamine + 0.01 $\underline{M}$ Calcium chloride, pH 7.3 (DTPA)
- Ammonium bicarbonate-DTPA (AB-DTPA)

The DTPA-extraction method, originally developed by Lindsay and Norvell (1978), is suitable not only for determining available zinc in calcareous soils but also for simultaneous determination of available Fe, Mn and Cu, because it provides for simultaneous complexing of all the four cationic micronutrients. This is ensured by adjusting the pH of the extracting medium (0.005 $\underline{M}$ DTPA, 0.1 M Triethanolamine and 0.01 $\underline{M}$ CaCl$_2$) to 7.3 with dilute HCl. Inclusion of CaCl$_2$ in the extractant enables it to attain equilibrium, minimizing the reaction with free carbonates in calcareous soils. The details of the DTPA extraction procedure have been carefully worked out (Liang and Karamanos, 1993). For reproducible results, it is essential to stick to the procedure. The DTPA-extraction method finds wide use since it is easy to follow and economical because it allows simultaneous determination of Fe, Mn, Cu and Zn in a single extract. Its suitability for evaluation of plant available zinc has been tested on a global scale (Sillanpaa, 1982). In a recent study, Abreu et al. (2002) made a comparison of four micronutrient extractants—Mehlich 1, Mehlich 3, DTPA and AB-DTPA—for evaluating available zinc in the soils of Brazil. All the four extractants were effective but DTPA was most suited as a measure of bioavailable zinc. On most alkaline soils, where zinc-deficiency is most prevalent, there is a highly significant correlation between DTPA extractable soil zinc and plant tissue zinc. The critical concentrations of zinc for delineation of zinc deficient soils may vary with the soil type and the crop cultivar, but on a broad basis soils with less than 0.6 μg DTPA extractable Zn g^{-1} soil are reported to be responsive to zinc amendment. This limit has been widely used for delineating zinc deficiency in different agro-climatic zones of India (Singh 2001).

A global study coordinated by Sillanpaa (1982) shows that pH corrected AAAc-EDTA extractable zinc (Lakanen and Ervio, 1971) may also be suitable for assessing zinc availability in soils. The AB-DTPA extraction procedure developed by Soltanpour and Schwab (1977) for simultaneous extraction of micronutrient cations has still not been used widely, possibly because it needs greater control on conditions of the extraction procedure than the DTPA-TEA procedure.

10.4.5 Molybdenum

Several soil tests have been developed to assess the availability of molybdenum in soils (Table 10.8). The suitability of the different soil test methods for determining plant available molybdenum has been discussed by Sims and Eivazi (1997) and Sharma and Chatterjee (1997).

Table 10.8. Common extractants for determining molybdenum availability in soils

- 1 $\underline{M}$ Neutral Ammonium Acetate
- 0.275 $\underline{M}$ Acid Ammonium Oxalate, pH 3.3
- Water
- Hot water (continuous leaching)
- Anion-exchange resin
- 1 $\underline{M}$ Ammonium Carbonate
- Ammonium Acetate-EDTA
- Ammonium Bicarbonate-DTPA

The most widely used extractant for nearly 50 years has been Acid Ammonium Oxalate (AAO), developed by Grigg (1953) for New Zealand soils. Contradictory observations have, however, been made about its suitability to delineate molybdenum deficiency (Mortvedt and Anderson 1982, references therein). Even Grigg (1960), who earlier developed the AAO test also found it unreliable, essentially because AAO solubilizes a part of the non-labile (Fe bound) pool of soil molybdenum. The suitability of AAO extraction has, however, been reported to increase if pH is considered as a factor (Cox, 1987). Extraction with water (Gupta and Mackay, 1965) is unsuitable because it solubilizes molybdenum in concentrations that, except in the alkaline soils, are too low to be analyzed colorimetrically without pre-concentration or else requires flameless (Graphite furnace) Atomic Absorption Spectrometry, facilities for which are available only in select laboratories.

A comparison of the AAO and the AB-DTPA extraction methods for available molybdenum show the latter to be more suitable (Pierzynski and Jacobs, 1986). The AB-DTPA extractable molybdenum is well correlated with plant molybdenum on alkaline soils and mine spoils (Wang et al. 1994). Use of AB-DTPA extraction method for alkaline soils offers added advantage of simultaneous extraction of the available forms of the cationic micronutrients (Fe, Mn, Cu, Zn), which may be limiting in these soils. The AB-DTPA extraction followed by direct estimation of molybdenum along with the cationic micronutrients using ICP spectrometry seems to be a promising proposition (Soltanpour, 1991). Sims and Eivazi (1997) have outlined an Anion-Exchange resin extraction procedure suited for acid soils.

So far, none of the chemical extractants has, however, been found satisfactory for developing critical deficiency limits for predicting molybdenum response. Perhaps a serious effort has also not been made in this direction because areas of molybdenum deficiency are localized and the deficiency can generally be effectively corrected by liming the soil.

Bioassay for Available Molybdenum

Sensitivity of the ascomycetaceous fungus *Aspergillus niger* to molybdenum nutrition (Mulder, 1939) formed the basis of a sensitive bioassay of molybdenum in soils (Nicholas and Fielding, 1951). The method was found suitable for determining the availability of molybdenum in New Zealand soils (Thornton, 1953). The bioassay procedure is suitable for determining molybdenum availability in alkaline soils (Sharma and Chatterjee, 1997). Details of the procedure, as developed by Agarwala et al. (1986), are described by Sharma and Chatterjee (1997).

10.4.6 Boron

A number of soil tests have been developed for determining boron availability in soils (Table 10.9). The first test developed for the purpose involved extraction with hot water by refluxing a mixture of boiling water and soil (Berger and Truog, 1939). In the form developed, or modified subsequently (Bingham 1982 and references therein), hot water extraction of boron still continues to be the most widely used and suitable method for determining boron availability in soils. Generally, less than 0.2 μg hot water-extractable boron g^{-1} soil is taken as the limit for deficiency (Adriano, 1986).

Extraction with dilute (0.01 or 0.02 $\underline{M}$) solution of calcium chloride (Jeffrey and Mc Callum, 1988; Parker and Gardener, 1981; Aitken et al, 1987) as such or with added mannitol (Cartwright et al. 1983) in about the same proportion as hot water but has the added advantage in that the extract is almost colorless and does not require discoloration (with charcoal), which may cause adsorptive loss of boron. Suitability of chelates (e.g. AB-DTPA) as extractants in acid soils, where boron deficiency is most prevalent, is questionable.

Table 10.9. Common extractants for determining boron availability in soils

- Hot water
- Mannitol
- 0.05 $\underline{M}$ Hydrochloric acid
- 0.02 $\underline{M}$ Calcium chloride
- 0.01 $\underline{M}$ Calcium chloride + 0.5 $\underline{M}$ Mannitol
- 1. $\underline{N}$ Ammonium acetate
- Sodium acetate-acetic acid (Morgan's reagent)

10.4.7 Chlorine

Water-soluble chlorides serve as a suitable measure of chloride availability in soils (Johnson and Fixen, 1990). Less than 2 μg Cl g^{-1} soil is considered deficient for plants. Such low concentrations are, however, seldom encountered, except in highly leached sandy soils.

10.5 BIOCHEMICAL AND OTHER TESTS

Several biochemical, structural and functional changes induced in response to micronutrient deficiencies have formed the basis of diagnosis of micronutrient deficiencies. Most common and widely used of these methods are alterations (decrease or increase) in enzyme activities. The idea of using enzyme activities as indicators of micronutrient status of plants was first mooted by Brown and Hendricks (1952). They suggested that decrease in catalase and peroxidase activities could provide an indication of limitation in iron supply and that of ascorbic acid oxidase activity as an indication of copper nutrient status of plants. The use of enzyme activities for diagnosing micronutrient deficiencies is based on known role of micronutrients as cofactors, activators or inhibitors of certain enzymes and the dependence of their activities on the tissue concentration of the constituent metal.

There are two ways in which enzyme activities could be used to diagnose micronutrient deficiencies. In the first method, the activity of the enzyme is measured in leaves showing visible symptoms of deficiency and compared with the activity in normal leaves of the same plant or comparable leaves of a healthy plant. In the other method, the leaves suspected of deficiency of a micronutrient are excised and infiltrated with that micronutrient and the enzyme activity is followed over a period of time. The rate of the increase in the activity of the enzyme indicates the severity of the deficiency. More severe the deficiency, more rapid is the induction of enzyme activity.

Besides enzyme activities, several other changes induced in response to micronutrient deficiencies have been suggested to be useful in diagnosing the deficiencies. Bussler (1981a) investigated the use of microscopic examination of plant tissues as an aid for diagnosis of micronutrient deficiencies and found reduced lignification of cells to be associated with copper deficiency. Abadia et al. (1991) described the usefulness of changes in photosynthetic pigments for characterization and detection of iron deficiency. Kriedmann and Anderson (1988) found manganese deficiency induced changes in chlorophyll fluorescence of wheat leaves to be indicative of manganese status of plants. Leaf reflectance and fluorescence characteristics may help early detection of other micronutrient deficiencies provided the spectral pattern typical to micronutrient deficiencies are characterized. With the availability of such a spectral database, the change induced by nutrient deficiencies can be discriminated. Adams et al. (2000) performed discriminate analysis on selected reflectance and fluorescence parameters for deficiencies of Mn, Cu, Zn and Fe in soyabean leaves. On the basis of discriminate analysis of chosen spectral parameters, 90% of Cu-deficient samples and 60% of the Mn-deficient samples were diagnosed. The percentage of samples detected deficient in zinc and iron on these bases was much less.

10.5.1 Iron Deficiency

Because of the limitation imposed by lack of correlation between the level of iron supply and the tissue concentration of iron, a necessity had been felt for developing biochemical tests that could give an idea of the physiologically active or functional iron in plants. Several biochemical methods have been developed for this. The relative suitability of such methods has been discussed by Chaney (1984). There is close correlation between the activity of catalase and often peroxidase (Sijmons et al. 1985) with the iron status of plants. The activities of these enzymes are also reduced under adverse soil conditions such as lime-induced chlorosis. Bar-Akiva (1984) explored the use of several substrates of peroxidase to identify the one best suited for its assay for rapid diagnosis of iron deficiency. Hsu and Miller (1968) studied the relationship between iron supply and the activity of the non-haem iron enzyme aconitase in soybean and found aconitase activity to be a suitable biochemical parameter for diagnosing iron deficiency. Mehrotra and Jain (1996) examined the activities of five iron enzymes in response to iron deficiency in sand-cultured radish, cauliflower and cabbage and found aconitate hydratase activity to be related to each iron supply, iron concentration and chlorophyll content. They found decrease in the enzyme activity even before iron deficiency effect on chlorophyll (chlorosis) became apparent. Abadia et al. (1991) have discussed the possibilities of using the concentrations of photosynthetic pigments as a measure of iron nutrient status.

10.5.2 Manganese Deficiency

Increase in peroxidase activity (Bar-Akiva, 1967) and decrease in phenylalanine ammonia lyase activity (Engelsma, 1972) under conditions of manganese deficiency led to the suggestion of their use in diagnosis of manganese deficiency. Leidi et al. (1987a) suggested the use of Mn-SOD activity as a marker of manganese deficiency. At the same time, there are reports of increase in Mn-SOD activity in response to manganese deficiency (Yu and Rangel, 1999). Kriedmann and Anderson (1988) have suggested the use of change in chlorophyll fluorescence as indication of manganese deficiency.

10.5.3 Copper Deficiency

Brown and Hendricks (1952) suggested that the activity of copper enzymes could be used for diagnosing copper deficiency. Brown (1953) showed that irrespective of whether the plants developed visible symptoms of copper deficiency or not, ascorbic acid oxidase activity was related to the level of copper supply. Bar-Akiva et al. (1969) used ascorbic acid (ascorbate) oxidase to diagnose copper deficiency in citrus trees. They infiltrated 0.2% solution of $CuSO_4$ in the leaves of plants suspected of copper deficiency and followed the induction of ascorbic acid oxidase activity. The increase in the enzyme activity was inversely related to leaf copper content. Several other studies showed the dependence of ascorbate oxidase activity on copper status of plants (Hewitt and Tatham, 1960; Perumal and Beattie, 1966; Loneragan et al. 1982; Delhaize et al. 1982, 1985). Based on the observed correlation between ascorbic acid oxidase activity and tissue concentrations of copper, Delhaize et al. (1982) described assay of ascorbate oxidase activity as a field test for diagnosis of copper deficiency. Nautiyal et al. (1999) reported close correspondence between the level of copper supply to rice plants and the leaf ascorbate oxidase activity. Delhaize et al. (1986) have suggested assay of copper diamine oxidase in newly emerging leaves as a measure of copper status of plants.

Copper-deficient plants show very poor lignification and microscopic examination for tissue lignification could give a fair idea of copper status of plants (Bussler, 1981b).

10.5.4 Zinc Deficiency

Wood and Sibley (1952) showed the usefulness of carbonic anhydrase (CA) activity as a measure of zinc status of plants. Subsequently, several workers found CA to be a suitable biochemical parameter for evaluation of zinc nutrient status of plants (Bar-Akiva and Lavon, 1969; Randall and Bouma, 1973; Dwivedi and Randhawa, 1974; Sharma et al. 1981; Rengel, 1995a; Chatterjee et al. 1998; Pandey and Sharma, 1998b). Recently, it has been shown that CA activity is not only a measure of the total zinc nutrient status, but functions as suitable parameter of its biochemical utilization. Higher activity of CA in leaves of plant genotypes with little difference in leaf tissue concentration of zinc indicates higher genotypic efficiency of zinc utilization (Rengel, 1995a; Hacisalihoglu et al. 2003). Superoxide dismutase (Cu-Zn SOD) is yet another zinc enzyme, activity of which is highly correlated to zinc supply and can be used as a biochemical method of assessing zinc nutrient status (Cakmak et al. 199c; Pandey et al. 2002b). As in case of CA, Cu-Zn SOD activity is also related to Zn efficiency of genotypes (Cakmak, et al. 1997c; Yu et al. 1999; Hacisalihoglu et al. 2003).

There is no convincing evidence of zinc being a specific cofactor of higher plant aldolase (fructose 1,6, biphosphatase aldolase), yet its activity is severely inhibited by zinc deficiency and aldolase has been suggested as an index of zinc deficiency (O' Sullivan, 1970; Bar-Akiva et al. 1971). In *Citrullus* (Sharma et al. 1981a) and black gram (Pandey et al. 2002b), aldolase activity decreased with decrease in zinc supply from sufficiency to deficiency levels. Ribonuclease is another enzyme, whose activity has been used for diagnosis of zinc deficiency. Its activity is increased under zinc deficiency (Kessler and Monselise, 1959; Dwivedi and Takkar, 1974; Johnson and Simons, 1979; Sharma et al. 1981; Snir, 1983; Bisht et al. 2002b; Pandey et al. 2002b). It is so even before the development of external signs of zinc deficiency, which adds to the advantage of using ribonuclease activity for diagnosing the incipient deficiency of zinc (Dwivedi and Takkar, 1974).

10.5.5 Molybdenum Deficiency

Nitrate reductase, which has specific requirement for molybdenum in presence of nitrate, has been suggested to be useful in diagnosis of molybdenum deficiency. Shabed and Bar-Akiva (1967) tested molybdenum status of citrus leaves through assay of nitrate reductase (NR) activity. They infiltrated citrus leaf discs with solutions of molybdate and nitrate (KNO_3) and followed the reduction of nitrate to nitrite over a period of time. The amount of nitrite produced was taken as a measure of NR activity and the rapidity of nitrite production as a measure of the severity of molybdenum deficiency. The faster the reduction of nitrate, the more severe was the deficiency of molybdenum. This was confirmed by Randall (1969). Over the years, several factors other than molybdenum deficiency have, however, been shown to influence NR activity. This makes the use of NR activity as an index of molybdenum deficiency less dependable, unless the enzyme activity is determined under well defined conditions, which defeats the purpose of rapid diagnosis of the deficiency.

10.5.6 Limitations in Use of Biochemical Tests

The use of metalloenzymes as biochemical markers of micronutrient status has the advantage of high sensitivity and specificity, but they have a major limitation in that identical enzymatic changes may also be induced by factors other than the limiting micronutrient. For example, activity of catalase, a haem protein, is not only inhibited under iron deficiency but also under zinc deficiency (Bisht et al. 2002a). Yu and Rengel (1999a) examined the suitability of metal-specific superoxide dismutases as a measure of plant status of the cofactor micronutrients. Their observations suggest that deficiency of one micronutrient (Zn, Cu, Mn or Fe) may not only affect the activities of the SOD containing that micronutrient as a cofactor but also other SOD forms. Cross talk between nutrient and other stresses may also complicate the enzymatic evaluation of the micronutrient status of plants. Activities of several enzymes suggested as suitable indicators of micronutrient status of plants are known to change in response to or as an adaptation to drought or oxidative stress. There are examples of enhancement in activities of SOD in response to deficiency of micronutrients that are SOD cofactors. Exposure of narrow-leafed lupins to manganese deficient nutrition reduced Mn concentration yet enhanced Mn SOD activity (Yu and Rengel, 1999b). Even otherwise, lack of a well-defined control, with which to make comparison, the precision required in assay of enzyme activities and the laboratory requirements for the tests do not favour the use of biochemical tests as a routine method for diagnosis of micronutrient deficiencies. These tests are, however, very useful in determining the functional fraction of the total accumulated micronutrients, or the efficiency of micronutrient utilization.

10.6 CROP RESPONSE

Crop response forms the ultimate measure of micronutrient deficiency and a dependable method for control of the deficiencies. Response of a crop to a nutrient amendment is a function of crop nutrient status, which in turn is determined by the bioavailability of the nutrient in soil. Crop yields obtained with different quantities of fertilizer/nutrient amendment offers an easy-to-follow and dependable method for evaluation of the nutrient status, absorbing the soil, plant and environmental influences.

Crop response to nutrient amendment is influenced by variations in chemical form of the fertilizer, the time and mode of its application, the plant genotype and the physiological stage of the plant at which fertilizer is applied. The response curves are, therefore, soil and plant specific. The theoretical and practical aspects of the response curves related to crop nutrient status are exhaustively discussed by Black (1993).

Besides field trials, crop response to micronutrients can also be tested through pot-culture trials under greenhouse conditions. In a pot trial, the soil to be tested for plant availability of a micronutrient is filled in pots of a suitable size and the particular micronutrient is added to the soil in the form of a suitable carrier. The crop plant to be tested for the response is raised in pots with graded doses of the micronutrients and quantified for biomass and/or harvest yield. The yield increment, relative to the unfertilized soil (control) indicates the response to the micronutrient, which is related to its deficiency in the soil and serves as a guide for optimizing crop yield through fertilizer amendment. For example, findings of a pot trial conducted to evaluate a soil for Zn deficiency (unpublished) (Fig 10.3; Table 10.10) suggest that Zn-fertilization of the soil under study @ 10 kg h^{-1} ($\equiv$ 10 mg kg^{-1} soil) can be predicted to nearly double the yield of maize grown thereon.

Fig. 10.3. Response of maize (*Zea mays* L.) to zinc (ZnSO$_4$) fertilization of a low zinc soil: Arranged left to right are pots applied zinc (ZnSO$_4$) @ 0(Control), 2, 5, 10, 15 and 20 mg kg^{-1} soil.

Table 10.10. Response of maize (*Zea mays* L.) to zinc ($ZnSO_4$) fertilization of a low Zn sandy loam soil in pot culture

Zn-application (mg kg⁻¹soil)	Available soil-Zn concentration (μg g⁻¹ dry wt)	Dry matter yield (g pot⁻¹)	Tissue Zn (μg g⁻¹ dry wt)	Yield response (% control)
0 (control)	0.28	62.8	16.4	-
2	0.48	80.8	21.6	29
5	0.56	96.3	24.8	53
10	0.86	121.4	38.0	93
15	0.90	100.4	60.1	60
20	2.08	97.3	85.5	55

Limited soil volume, root space and nutrient availability in the rhizosphere make pot-culture conditions different from field conditions (Huang et al. 1996), but for short duration, annual crops, it still serves a useful purpose in evaluating soil micronutrient status and fertilizer use for optimizing crop yield.

Plate 1. Iron deficiency symptoms in pea (top left), flax (top right), cowpea (middle, left), groundnut (middle, right), pearl millet (bottom, left) and radish (bottom, right)

Plate 2. Manganese deficiency symptoms in mustard (top, left), soybean (top, right), green gram (bottom, left) and papaya (bottom, right)

Plate 3. Copper deficiency symptoms in chickpea (top, left), flax (top, right), barley (bottom, left) and mango (bottom, right)

Plate 4. Zinc deficiency symptoms in cowpea (top, left), cucumber (top, right), wheat (bottom, left) and cotton (bottom, right)

Plate 5. Molybdenum deficiency symptoms in mustard (top, left), sunflower (top, right), aubergine (bottom, left) and cabbage (bottom, right)

Plate 6. Boron deficiency symptoms in maize (top, left), cucumber (top, right), sunflower (middle, left), pearlmillet (middle, right), cowpea (bottom, left) and pigeon pea (bottom, right)

CHAPTER 11

AMELIORATION OF DEFICIENCIES

11.1 GENERAL

Amelioration of micronutrient deficiencies involves two approaches. The first approach aims at correction of deficiencies through application of the limiting micronutrients or amendments that increase their availability to plants. The second approach aims at raising crop genotypes with high nutrient efficiency on soils lacking in their availability. Once a micronutrient has been identified to be deficient in a soil and the extent of deficiency has been evaluated by appropriate methods (Chapter 10), it can be applied to the soil (soil amendment) or the plant (foliar amendment) to overcome the deficiency. This approach has been followed since the beginning of cultivation by man and is still the most widely followed method for amelioration of micronutrient deficiencies (Murphy and Walsh, 1991). Where the availability of micronutrients is limited by some soil chemical condition (e.g. high pH), deficiencies can be ameliorated through soil amendments (e.g. liming), causing an increase in the availability of micronutrients through modification of the soil chemical conditions. Micronutrient deficiencies caused by excessive salinity can be ameliorated by application of gypsum ($CaSO_4.7H_2O$), which increases exchangeable Ca^{2+} and decreases exchangeable Al (Shainberg et al. 1989). Availability of micronutrients (Fe, Zn) can also be increased through biological methods such as VAM associations or microbial siderophores. Rhizosphere microorganisms can affect plant nutrition by influencing the availability of plant nutrients, growth and morphology of roots and nutrient uptake processes (Rovira et al. 1983). Organic manures provide another potential source of micronutrients. They have an edge over the artificial fertilizers in that they also improve soil conditions to favour enhanced availability of native as well as applied micronutrients.

The second approach, that finds favour over the first, because of cost effectiveness, energy costs and sustainability, is the plant-based approach wherein the deficiency constraint is overcome by going in for cultivation of plant genotypes that, by virtue of high nutrient efficiency, can perform well on soils of low micronutrient availability. Use of genotypes, that have high efficiency of absorption, translocation and/or utilization of micronutrients, is particularly advantageous in overcoming deficiencies imposed due to adverse soil conditions such as calcareousness, salinity and sodicity, which are otherwise difficult to correct. During the recent years, greater thrust has been laid on overcoming deficiencies (and toxicities) of micronutrients through use of efficient genotypes (Sattelmacher et al. 1994; Welch, 1995; Graham and Welch, 1996; Pearson and Rengel, 1997; Rengel, 1999; Nielsen and Jensen, 1999). The choice of the approach for management of deficiencies differs with particular micronutrients, soil conditions and crop species.

11.2 MICRONUTRIENT AMENDMENTS

The limiting micronutrient(s) can be supplemented in different ways. This may be achieved by application of the limiting micronutrient in a suitable, inorganic or organically complexed form as (a) soil amendment, (b) foliar application or through (c) seed treatment.

Efficiency of soil application of a micronutrient for correction of deficiency depends on the proportion in which the added fertilizer contributes to the labile pool of micronutrients in the rhizosphere. This is influenced to varying extents by the interactions of the added micronutrient carriers with the soil physico-chemical conditions and environmental influences. A large proportion of boron and manganese added to the soil through inorganic carriers may be rendered unavailable, particularly in high pH soils, and this, reduces their efficiency as corrective measures. The effectiveness of foliar application of micronutrients, on the other hand, depends on their phloem mobility. High phloem mobility of molybdenum makes its foliar sprays an effective method for correction of molybdenum deficiency. Seed micronutrient content has an important bearing on seed vigour and viability. When sown to soils deficient in a micronutrient, seeds rich in that micronutrient can establish and support early seedling growth much better than seeds lacking in that micronutrient. Enrichment of seeds in micronutrients either through parental nutrition or seed treatment have been found effective in overcoming their marginal deficiencies. The effects of seed coatings and seed treatments on the establishment of seedlings on nutritionally poor soils are discussed by Scott (1989).

11.2.1 Iron Deficiency

SOIL AMENDMENT

Several iron salts find use in correction of iron deficiency (Table 11.1), but they have a common drawback in that a large proportion of iron supplied through the inorganic carriers is rendered unavailable, and this necessities their application at rates that are prohibitive. Amongst the inorganic iron fertilizers, the most commonly used are the ferrous or ferric sulphate. Both are supplied at rates providing 100 to 500 kg Fe ha^{-1}.

Table 11.1. Common inorganic iron fertilizers

Fertilizer	Chemical formula	Fe content (%)
Ferrous sulphate	$FeSO_4.7H_2O$	20.5
Ferric sulphate	$Fe_2(SO_4)_3$	20.0
Ferrous ammonium sulphate	$(NH_4)_2 SO_4, FeSO_4.6H_2O$	14.0
Ferrous carbonate	$FeCO_3$	42.0

The problem of non-availability of iron supplied through inorganic fertilizers can be circumvented through use of synthetic iron chelates (Table 11.2).

Table 11.2. Common synthetic iron chelates

Iron-chelate	Common name
Ferric-Ethylene diamine tetraacetic acid	Fe-EDTA
Ferric-Diethylene triamine pentaacetic acid	Fe-DTPA
Ferric-Hydroxyethyl ethylene diamine tetraacetic acid	Fe-HEEDTA
Ferric-Cyclohexane trans 1,2-diamino tetraacetic acid	Fe-CDTA
Ferric-Ethylene diamine-dihydroxy phenylacetic acid	Fe-EDDHA

Synthetic chelates of iron, in particular Fe-EDDHA (6% Fe), have been successfully used in correcting iron deficiency. The rates of application of the chelates vary widely. Even at low concentrations (0.5 to 1.0 kg ha^{-1}), they provide effective control of iron deficiency. The only deterrent in their large-scale use is their high cost. Complexes of iron with naturally occurring organic compounds, such as lignin sulphonates, polyflavonoids and methoxyphenyl propane, have also been used for correction of iron deficiency. Use of iron frits, wherein iron is fused to glass, serve as a source of slow releasing iron. Their use has been advocated for correction of iron deficiency on acid soils with high percolation rates.

Because of widespread occurrence of iron chlorosis, particularly on calcareous soils, its prevention and correction have always been areas of major interest in plant nutrition (Hogstrom, 1984; Mortvedt, 1991b; Wallace, 1991, and references therein). Besides inorganic iron fertilizers and iron chelates that have been in wide use for several decades, recommendations have been made for the use of soil application of acid-and Fe-fortified organic wastes and sulphur-pyrite mixes (Wallace, 1991). Effectiveness of Fe chelates is increased by making them slow release or applying with seed (Wallace, 1992).

FOLIAR APPLICATION

Foliar sprays of both inorganic iron compounds and iron chelates find wide application in correction of iron deficiency, particularly so in fruit and vegetable crops. The most widely used inorganic iron fertilizer for spray application is ferrous sulphate. Foliar spray of 1 to 3% solution of $FeSO_4.7H_2O$ at 300-400 l h^{-1} is generally effective in correction of iron chlorosis. Wide use is made of foliar application of iron chelates for correction of iron deficiency in calcareous soils. Effective recovery from iron chlorosis is commonly obtained with a high volume spray (500 to 1,000 l ha^{-1}) of 0.05% solution of Fe-EDTA repeated at fortnightly intervals.

SOIL MANAGEMENT

Iron deficiency on high pH (including calcareous) soils can be successfully prevented by soil acidification through sulphur compounds. This is, however, not specific for iron deficiency and involves high expenses.

11.2.2 Manganese Deficiency

SOIL AMENDMENT

Several inorganic compounds of manganese are used for correction of manganese deficiency. The most widely used is manganese sulphate ($MnSO_4.3 H_2O$). Its rate of application may vary from 5 to 100 kg Mn ha^{-1}, depending on the severity of deficiency, the soil pH and the method of application-broadcast or band placement. Broadcast is preferred for soils with acidic pH (< 6.5). When soil pH exceeds 6.5, band placement gives better results. In general, precipitation of manganese at high pH decreases the efficiency of manganese fertilizers on alkaline soils. Even otherwise, soil application of manganese is often not considered a dependable method for correction of manganese deficiency.

FOLIAR APPLICATION

Usual method for control of manganese deficiency is through foliar application of manganese sulphate (Reuter et al. 1988). Foliar application is preferred over soil application because of relatively low rates of application and low cost involvement. But, because of poor residual value, control of manganese deficiency through foliar spray is obtained only on repeat applications of manganese sulphate (at a concentration of 10 kg ha^{-1}). Foliar application of chelated manganese (Mn-EDTA) provides another effective method for correction of manganese deficiency but on economic considerations its use is limited. Foliar spray of manganese containing dithiocarbamates, used for control of pathogenic diseases, are reported to add to manganese content of the foliage and reduce the severity of manganese deficiency.

SEED TREATMENT

Manganese deficiency can be substantially prevented by use of seeds of high manganese content. Such seeds can be obtained from parent plants that had been adequately fertilized with manganese or by soaking the seeds in a solution of manganese sulphate. Longnecker et al. (1991b) observed improvement in plant growth and grain yield of barley grown on manganese-deficient soils when the crop was raised from seeds of high manganese content. Wheat plants raised from seeds of high manganese content, showed low incidence of take-all than plants raised from seeds of relatively low manganese content (Huber and McCay-Buis, 1993; McCay-Buis et al. 1995; Pedler, 1996). It is suggested that use of seeds with high manganese may provide more effective control of manganese deficiency in wheat than manganese fertilization of soil (Moussavi-Nik at al. 1997).

11.2.3 Copper Deficiency

SOIL AMENDMENT

Effective and persistent control of copper deficiency is secured through application of copper sulphate, copper oxide, Cu-chelates (Table 11.3) and copper containing slag. The most widely used and perhaps the cheapest copper fertilizer is copper sulphate. The most common dose of copper application to soil through copper sulphate is 100g Cu ha^{-1}.

Table 11.3. Common copper fertilizers

Fertilizer	Chemical formula	Cu content (%)
Copper sulphate	$CuSO_4.5H_2O$	25
	$CuSO_4.H_2O$	35
Copper oxide	Cu_2O	89
	CuO	75
Copper chelates	Na_2-Cu EDTA	13
	Na-Cu EDTA	9

Copper chelates are easily available but they are expensive and provide little advantage over inorganic copper compounds. The most widely used dose of chelated copper (Cu-EDTA) is 30g Cu ha^{-1}. The rate of copper application has to be based on the severity of copper deficiency and the method of application. In both inorganic and chelated form, relatively large doses are required for broadcast application than for band placement. Since copper fertilizers have a high residual effect, they provide a long-term solution to copper deficiency. Soil application of inorganic copper salts, oxides or slow release compounds @ 15 kg Cu ha^{-1}, has been reported to provide effective protection against copper deficiency for over a decade. Care should, however, be exercised in selecting the application rates of copper fertilizers to ensure that they do not build up copper concentrations in soil to toxic limits.

FOLIAR APPLICATION

Foliar application of copper in the form of inorganic salts, oxides or chelates provides effective correction of copper deficiency. This is preferred to soil application as it avoids build up of Cu in soils to toxic limits. Most widely used source of copper for foliar application is copper sulphate. Neutralization of copper sulphate with lime prevents the toxic effects on the foliage. The need for the treatment has to be based on soil analysis and/or copper deficiency symptoms in the previous crops. Generally, a single spray is effective in correcting even severe deficiency. Copper chelates of EDTA or HEEDTA form other effective methods for control of copper deficiency but repeated applications of these may be required for optimum results.

11.2.4 Zinc Deficiency

SOIL AMENDMENT

Soil application of zinc-containing inorganic fertilizers offers effective control of zinc deficiency. Several salts of zinc are used for the purpose (Table 11.4).

Table 11.4. Common inorganic zinc fertilizers

Fertilizer	Chemical formula	Zn content (%)
Zinc sulphate	$ZnSO_4.7H_2O$	23
	$ZnSO_4.H_2O$	36
Zinc oxide	ZnO	60–80
Zinc chloride	$ZnCl_2$	45–52
Zn carbonate	$ZnCO_3$	56
Zinc oxide-sulphate	$ZnO-ZnSO_4$	55
Zinc ammonium phosphate	$Zn(NH_4)PO_4$	37
Sulpherilite	ZnS	60

The most widely used zinc fertilizer is zinc sulphate. Generally, zinc sulphate is applied at a rate providing 5 to 20 kg Zn ha^{-1}. Actual rates would depend on the crops to be fertilized, the type of soil and the method of application. Metallic zinc (as zinc dust) is also used as a soil amendment.

The efficiency of zinc fertilizers is influenced by several environmental and biological factors. Morgan (1980) described large variations in flux of zinc (and phosphorus) fertilizers with changes in soil temperature. Nayyar and Takkar (1980) evaluated the suitability of different sources of zinc for fertilizing rice on alkali soils. According to them, zinc deficiency on calcareous soils can be readily corrected by soil application of inorganic zinc salts such as $ZnSO_4$. Effective control of zinc deficiency is secured through soil application of zinc chelates (Zn-EDTA, Zn-HEDTA) at considerably low (0.5 to 1.0 kg Zn ha^{-1}) concentrations.

FOLIAR APPLICATION

Foliar application of zinc is often the preferred form of zinc fertilization of fruit trees and vegetable crops. It is also the method of choice where adverse soil chemical conditions render a large proportion of soil applied zinc unavailable to plants. The most commonly used inorganic carrier of zinc for foliar application is zinc sulphate (neutralized with lime). Effective control of zinc deficiency is secured through 2 or 3 foliar applications of 0.5% $ZnSO_4$ + 0.25% lime sprayed @ 400 l ha^{-1}. Use of a wetting agent improves the efficiency of the treatment. Haslett et al. (2000) have shown that foliar application of zinc, either in inorganic or organic form, can effectively meet the zinc requirements of wheat.

SEED TREATMENT

Seed treatment of zinc may not provide total correction of zinc deficiency. It has, however, been shown that zinc content of seeds contributes to early seedling growth and that high zinc content of seeds may act as a starter-fertilizer, helping establishment of seedlings on low zinc soils (Rengel and Graham, 1995a,b; Grewal and Graham, 1997).

11.2.5 Molybdenum Deficiency

SOIL APPLICATION

Molybdenum deficiency can be successfully ameliorated through use of inorganic molybdenum fertilizers (Table 11.6). Dose of molybdenum fertilizers to correct the deficiency has to be based on the molybdenum requirement of plants, which differs in nodulating and non-nodulating plants and is influenced by the level and form of nitrogen fertilization. Studies on soybean have shown that response to molybdenum application is largely restricted to nodulating plants (Parker and Harris, 1977). This is attributed to greater need of molybdenum for nitrogen fixation than for assimilation of nitrate. As would be expected on the basis of need of molybdenum for nitrate reduction (Hewitt and McCredy, 1956), the response to molybdenum application is high when plants are fertilized with nitrogen as nitrate than with ammonium fertilizers. The level of nitrogen fertilization also influences response to molybdenum application. Response is high when soils are not adequately supplied with nitrogen. Sources and methods of molybdenum fertilization of high molybdenum requiring crops are discussed by Mortvedt (1997). Application of molybdenum at low rates along with inoculation of suitable strains of *Rhizobium* may substitute for high rates of molybdenum fertilization. Generally, the fertilizers are applied at a rate delivering 70 to 200 g Mo ha^{-1}. Crops having higher molybdenum requirement may need higher rates of fertilization (up to 400 g Mo ha^{-1}).

Table 11.5. Common molybdenum fertilizers

Fertilizer	Chemical formula	Mo content (%)
Sodium molybdate	$NaMoO_4.2HO$	39
Ammonium molybdate	$(NH_4)_6 Mo_7O_{24}.4H_2O$	54
Molybdenum trioxide	MoO_3	66
Molybdenite	MoS_2	60

FOLIAR APPLICATION

As molybdenum is phloem mobile (Kannan and Ramani, 1978; Brodrick and Giller, 1991), most crops that develop molybdenum deficiency benefit from foliar sprays of molybdenum (Gupta and Lipsett, 1981). Generally, foliar application of 0.1 or 0.3% solution of ammonium or sodium molybdate produces good results. Foliar application of molybdenum is very effective in increasing the yield of legumes (Adams, 1997), because it is preferentially translocated to the root nodules (Brodrick and Giller, 1991) and this benefits both N_2 fixation and final yield of the host plants. In view of molybdenum requirement for plant reproductive development foliar application of molybdenum at the onset of reproductive phase is expected to produces better results.

SEED TREATMENT

Seed treatment provides an effective and economic method for prevention of molybdenum deficiency. Seed treatment has an advantage over soil application because the low quantities in which molybdenum is required by plants can be better administered through seed treatment than soil application. Gurley and Giddens (1969) have shown a direct relationship between molybdenum content of seeds and subsequent seed yield of soybean grown on low molybdenum soils.

Seed application of molybdenum generally involves palleting of seeds with molybdenum trioxide (MoO_3). The method has been found effective in overcoming molybdenum deficiency in tropical pasture legumes (Kerridge et al. 1973). The rate of application through seed treatment ranges between 50 to 100 g Mo ha^{-1}. Hafner, et al. (1992) have shown that in certain legumes, such as groundnut, seed palleting with 100 g molybdenum produces higher increments in dry matter yield and plant nitrogen content than soil application of 60 kg ha^{-1} of mineral fertilizer nitrogen.

SOIL MANAGEMENT

Liming of acid soils (to pH around 6.5) provides an effective method for correction of molybdenum deficiency. Effective and long-term control of molybdenum deficiency through liming has to be based on knowledge of soil acidity as well as molybdenum status. Excessive application of lime may cause adsorption of soil molybdenum to $CaCO_3$, reducing its availability to plants. On soils that are low in molybdenum and acidic in reaction, the best and longest lasting response may be provided by a combination of liming and molybdenum fertilization of soils (Adams et al. 1990).

11.2.6 Boron Deficiency

SOIL AMENDMENT

Several boron compounds are used as carriers of boron. It is commonly applied as borax (disodium tetraborate) or as soluble borates, available under proprietary names (Table 11.6).

Table 11.6 Common boron fertilizers

Fertilizer	Chemical formula	B content (%)
Borax	$Na_2B_4O_7.10H_2O$	11
Boric acid	H_3BO_3	17
Sodium tetraborate		
Fertilizer borate 46	$Na_2B_4O_7.5H_2O$	14
Fertilizer borate 65	$N_2B_4O_7$	20
Sodium pentaborate	$Na_2B_{10}O_{16}.10H_2O$	18
Solubor	$Na_2B_4O_7.5H_2O$ + $Na_2B_{10}O_{16}.10\ H_2O$	20–21
Colemanite	$Ca_2B_6O_{11}.5H_2O$	10

Because of wide differences in plant requirement for boron and the low concentration at which boron may become phytotoxic, the rates of boron application have to be chosen carefully. Application at a rate of 1 to 2 kg B ha^{-1} (=10 to 20 kg borax ha^{-1}) may benefit most crops. High boron requiring crops such as sugarbeet may need more. Boron fertilizers may be applied either broadcast or as basal placement. In keeping with the low rates of application, borax or the other more soluble boronated compounds can also be applied as spray to the soil just before planting. As response to soil application of boron is not very persistent on sandy soils, correction of severe boron deficiency may necessitate repeat applications of boron fertilizers, but this has the risk of toxicity to susceptible crops. Boron frits, in which boron is contained in a moderately soluble glass, may serve as slow release boron carriers.

FOLIAR SPRAY

Foliar application of boron as boric acid and other boronated compounds provide effective control of boron deficiency. Good response is obtained both on vegetative growth and reproductive development. Schon and Blevins (1990) reported marked increase in the number of pod bearing branches as well as number of pods in soybean in response to foliar application of boron. Hanson (1991) reported a large increase in the number of flowers and fruits of sour cherry trees sprayed with boron. Repeated sprays of 0.5 to 1.0 kg B ha^{-1} (as borax, boric acid or sodium tetraborate) provide effective control of boron deficiency. Early sprays are more effective.

11.2.7 Chlorine Deficiency

Chlorine deficiency is generally not a field problem. There are, however, reports of benefits from Cl$^-$ fertilization. This relates to both increase in yield and resistance from pathogenic infections. Fertilization of crops with chloride fertilizers (40 kg Cl ha^{-1}) has been reported to prevent the incidence of take-all disease in the Pacific Northwest (Christensen et al. 1987).

11.3 USE OF MICRONUTRIENT EFFICIENT GENOTYPES

In general, the traditional method of applying micronutrients to plants subjected to their deficiency provides effective control of their deficiencies but this method involves high-energy inputs. Costs on both labour and fertilizers are high. Continuous fertilizer usage also contributes to resource depletion and, at times, soil pollution. The alternative approach of 'tailoring the plant to fit the soil' (Foy, 1983) circumvents these problems. Growing plant genotypes that can produce reasonably well under conditions of low micronutrient availability, where the other commonly cultivated crop genotypes fail, offers a low input and ecologically safe approach for amelioration of the deficiencies.

Ability to withstand or tolerate the deficiency stress is a function of the nutrient efficiency of the genotype, which is a heritable trait. High nutrient efficiency could be due to high ability for nutrient uptake (uptake efficiency), shoot to root transport (transport efficiency) and/or biochemical utilization (utilization efficiency). Considerable information has been gathered on genotypic differences in micronutrient efficiency of plants and the mechanisms which contribute to these differences (Viets, 1966; Lucas and Kezek, 1972). Different mechanisms contributing to nutrient efficiency of genotypes have been discussed by Sattelmacher et al. (1994) and reviewed by Rengel (1999b). In general, uptake efficiency is a function of root morphology, root exudations and nutrient transport across plasmalemma. Not only nutrient influx but efflux contributes to genotypic differences in nutrient efficiency. Rengel and Graham (1996) showed high efflux of zinc by Zn-inefficient wheat genotypes. Mechanisms contributing to nutrient efficiency may be constitutive or inducible (inductive). The latter are expressed by the efficient genotypes in response to deficiency stress, as is the case with iron deficiency stress. In Fe-efficient genotypes of both strategy I and strategy II plants, the efficiency mechanisms are induced only in response to Fe-deficiency stress (Jolley et al. 1996; Pearson and Rengel, 1997). Likewise, Zn-efficient genotypes of wheat are reported to show increased rates of Zn uptake only when subjected to Zn deficiency stress (Rengel and Wheal, 1997). Better nutrient efficiency could result from more efficient transport of nutrients from root to shoot involving their loading into the xylem sap, and in case of nutrients transported as organic complexes, the availability of the organic ligands. Genotypes also differ in their capacity to make functional use of the nutrients. It has been shown that for about the same concentration of Zn, as in Zn-inefficient genotypes of wheat, the Zn-efficient genotypes show higher activity of carbonic anhydrase (Rengel, 1995a), which may contribute to higher rates of photosynthesis (Fischer et al. 1997), and superoxide dismutase (Cakmak et al. 1997c), which functions as a component of the antioxidative defense system (Cakmak, 2000).

Notwithstanding the mechanisms contributing to higher nutrient efficiency, which may, except for isogenic lines, differ in different genotypes, the use of efficient genotypes provide an opportunity to manage the deficiencies unless they are very severe. Under conditions of severe deficiency, a combination of efficient genotypes and fertilizer amendment, at a much lower rate than needed for an inefficient genotype, may provide the desired yield response. Efficient genotypes are of particular advantage in management of deficiencies on problem soils, where their correction through fertilizer use alone is difficult.

In view of the fact that use of nutrient-efficient (= tolerant) genotypes provides a low-input and environmentally friendly approach for management of deficiencies (Lynch, 1998), increasing thrust is being laid on breeding and identifying the existing genotypes for nutrient efficiency. Nutrient efficiency traits being inheritable and efficiency being the dominant factor, development of nutrient efficient genotypes should be possible through conventional breeding (Cianzio, 1999). So far, major emphasis has, however, been laid on identifying existing crop cultivars for tolerance to micronutrient deficiencies and toxicities through appropriate screening methods.

IRON DEFICIENCY

Because of the several factors that render iron unavailable to the plant, iron deficiency forms the most common and widespread micronutrient disorder of plants, particularly on calcareous soils. Long-term, sustainable solution to the problem, through a soil-based approach is not a successful story. The problem finds a more practical, sustainable and cost-effective solution in raising crop cultivars or genotypes, which have high efficiency of uptake, transport and/or utilization of iron. Interest in genotypic approach for overcoming the constraints of lime-induced deficiency of iron dates six decades back, when Weiss (1943) showed iron chlorosis in soybean to be a heritable trait, determined by a single recessive gene pair. It was observed that when grown on calcareous soils, soybean cv. PI 54619-5-1 (PI) developed iron chlorosis but cv Hawkee (HA) showed apparently healthy growth (free from iron chlorosis) (Brown, 1961). The difference in the tolerance of the two cultivars to iron chlorosis was found to be related to the difference in the ability of their roots to absorb and translocate iron (Brown, 1963). Higher efficiency of iron uptake by HA roots was related to higher capacity of its roots to reduce Fe^{3+} to Fe^{2+}, in which form alone soybeans absorb iron. Over the years, it has been established that, under conditions of iron deficiency, transport of iron across the plasmalemma involves the activity of plasmalemma bound Fe(III) chelate reductase (Bienfait, 1985), and that it is the main factor contributing to iron efficiency of strategy I plants. When subjected to iron deficiency, roots of some dicotyledonons plants secrete riboflavin, which leads to enhanced uptake of iron (Welkie and Miller, 1993). The Fe-efficient genotypes have been shown to possess greater capacity for riboflavin secretion in response to Fe-deficiency than the Fe-inefficient genotypes (Jolley et al. 1991; Welkie, 1996).

The Fe-efficiency of strategy II genotypes is essentially a function of the capacity of their roots to produce and release phytosiderophores in response of iron deficiency stress (Römheld and Marschner, 1990; Jolley and Brown, 1991a) and the uptake kinetics of the Fe-phytosiderophores (Von Wiren et al. 1995). The phytosiderophores released by plant roots in response to iron deficiency from a complex with Fe^{3+} and contribute to increased uptake of Fe^{3+} in the chelated form. The capacity to produce and release the phytosiderophores is genetically determined. The Fe-efficient genotypes of the strategy II plants, that show tolerance to iron chlorosis on calcareous soils, have been found to be efficient in release of phytosiderophores such as *hydroxymugineic acid* (HMG). The efficiency of cereal crop species growing on calcareous soils is related to the capacity of their roots to release phytosiderophores. It has been suggested that release of phytosiderophores by roots can form a criterion for evaluation of Fe-efficiency of the graminaceous plants (Hansen et al., 1995, 1996). Such an attempt should, however, take into account the fact that release of phytosiderophores by roots shows diurnal variations (Römheld and Marschner, 1981a; Zhang et al. 1991b; Walter et al, 1995; Cakmak et al. 1998).

A prerequisite to management of iron-deficiency using the iron-efficient genotypes is the identification of such genotypes through suitable screening procedures. Identification of bicarbonate as the soil factor inducing chlorosis in susceptible genotypes of soybean grown on calcareous soils (Coulombe et al. 1984a) led to development of a screening method in which bicarbonate was included in the nutrient solution to simulate soil calcareousness (Graham et al. 1992). The method developed for screening soybean genotypes for iron effectivity included bicarbonate and low (1.5 to $2\mu m$) concentration of iron supplied as Fe-EDDHA (Coulombe et al. 1984b). Subsequently, Fe-EDDHA was replaced with Fe-DTPA as the source of iron (Coulombe and Chaney solution) because DTPA has higher iron buffering capacity than EDDHA (Chaney et al. 1989). The method was further improved for screening a wide range of dicots (strategy I plants) for Fe-chlorosis (Chaney et al. 1992a,b). While use of 25 μm Fe-DTPA was suggested for general screening, 15 μm Fe-DTPA was recommended for screening highly chlorosis resistant cultivars. Methods have also been developed for screening of plant genotypes for iron chlorosis by raising seedlings on a calcareous soil under greenhouse conditions. Ocumpaugh et al. (1992) found the method suitable for screening oats for iron deficiency. A positive correlation between root Fe^{3+}-reducing capacity and Fe-efficiency (Camp et al. 1987; Jolley and Brown, 1987; Tipton and Thawson, 1985) formed the basis for development of large scale short-term methods for screening of plant genotypes for iron efficiency (Jolley et al. 1992; Stevens et al. 1993). The Fe^{3+} reduction capacity of roots of soybean genotypes was found to be highly correlated to their field susceptibility rating for iron chlorosis (Jolley et al. 1992). Lin et al. (1998) made a comparative study of linkage map and quantitative traits loci (QTL) in $F_{2:4}$ lines from two soybean populations-Pride B 216 X A 15 and Anoka X A7 grown in nutrient solution and on calcareous soils under field conditions. Identical traits were expressed under both the conditions, which showed that both the systems identified similar genetic mechanisms of iron uptake and/or utilization.

MANGANESE DEFICIENCY

The general, mechanisms contributing to nutrient uptake, transport and utilization fail to provide a satisfactory basis for manganese efficiency of genotypes (Rengel, 1999). It is largely so because manganese efficiency is expressed in the soil system but not in solution cultures (Huang et al. 1994) that provide a suitable technique for investigating the physiological basis of genotypic differences in nutrient efficiency. According to Rengel (1999), the main factor that contributes to manganese efficiency of a genotype is the capacity of plant roots to excrete substances that facilitate larger mobilization of manganese from the rhizosphere. The root secretions possibly include reductants and organic ligands capable of chelating manganese and microbial stimulants. Secretions that favour the growth of the Mn-reducing organisms or inhibit the Mn-oxidizing microorganisms are likely to contribute to manganese efficiency (Timomin, 1965). Differences in the pattern of internal compartmentalization and remobilization of manganese are other likely factors contributing to genotypic differences in manganese efficiency (Huang et al. 1993).

Differences in tolerance of crop genotypes to manganese deficiency have been reported (Graham, 1988; Bansal et al. 1991). Based on field screening on low manganese soils, Bansal et al. (1991) have identified some manganese deficiency tolerant genotypes of wheat. Ingeneral, information on genotypic differences in crop tolerance to manganese deficiency is limited possibly because manganese deficiency does not pose a serious problem to crops on a large scale.

COPPER DEFICIENCY

Graham et al. (1987a) described copper efficiency 'as the ability of a genotype to yield well on a copper deficient soil, and also, taking account of different yield potentials, as relative yield of paired-Cu and +Cu plants i.e. (-Cu/+Cu) x 100'. Differences in Cu-efficiency of genotypes have been made out best in case of cereal crops, which are sensitive to copper deficiency (Agarwala et al. 1971; Graham et al. 1987a). Rye has high copper efficiency, which is attributed to location of gene(s) controlling copper efficiency on the long arm of its 5 R chromosome (5 RL) segment (Graham et al. 1987a). The efficiency of this segment has been successfully transferred to the rye-wheat hybrid *triticale* (Graham and Pearce, 1979) and, through back-crossing and selfing, into wheat (Graham et al. 1987a). The wheat-rice translocation lines show marked increase in copper efficiency. Their grain yield on copper deficient soils is reported to be more than double that of the parental lines.

ZINC DEFICIENCY

With the possible exception of iron, more studies have been carried out to study the causes for genotypic differences in zinc efficiency than any other micronutrient. The subject has been recently reviewed by Rengel (1999). There are indications that enhanced uptake of zinc by the Zn-efficient genotypes is an inducible response evoked in response to zinc deficiency (Rengel and Wheal, 1997). The Zn-efficient genotypes also show increased apoplast to symplast transport of zinc under conditions of Zn-deficiency (Rengel and Graham, 1996; Rengel and Howkesford, 1997; Rengel et al.

1998). Genotypic differences have also been shown to be influenced by rates of zinc efflux. Higher efflux of zinc by Zn-inefficient genotypes could result in decrease in zinc concentration (Rengel and Graham, 1996). Some genotypes may show higher Zn-efficiency because of a higher capacity for biochemical utilization of zinc. Even with little change in uptake or transport of zinc under zinc deficiency conditions, the Zn-efficient genotypes may show higher activities of zinc enzymes such as carbonic anhydrase (Rengel 1995a, Hacisalihoglu et al. 2003) and Cu/Zn superoxide dismutase (Yu et al. 1999; Hacisalihoglu et al. 2003). Hacisalihoglu et al. (2003) have shown that Zn-efficient and Zn-inefficient genotypes of wheat, which show no differences in uptake and transport of zinc and do not differ in water soluble zinc content or sub-cellular compartmentalization of zinc, show marked differences in the expression of zinc enzymes. When subjected to zinc-deficiency stress, the Zn-efficient genotype Kargiz, showed higher activities of Cu/Zn SOD and carbonic anhydrase than the Zn-inefficient genotype BDME. Northern analysis showed that the activities of the two enzymes in the Zn-efficient genotype were upregulated in response to Zn-deficiency stress. Thus, Zn efficiency of Kargiz is a function of its higher ability for biochemical utilization of zinc under conditions of its limited availability. Zinc-efficient genotypes of wheat are also reported to have high tolerance to crown rot decrease caused by *Fusarium graminiarium* (Grewal et al. 1996).

While little advance has been made in breeding zinc-efficient genotypes (Cianzio, 1999), a large number of crop species have been investigated for genotypic differences in zinc efficiency (Table 11.7). In India, where zinc deficiency forms a major constraint in crop production, promising varieties of Indian crop plants have been evaluated for tolerance to zinc deficiency (Takkar, 1993). Agarwala et al. (1978) examined 35 rice varieties for tolerance to zinc deficiency. Rice variety Sabarmati, showed high tolerance to zinc deficiency (Fig. 11.1). Screening of wheat varieties for tolerance to zinc deficiency showed var. WL 212 to be tolerant to zinc deficiency (Fig. 11.2) (Agarwala and Sharma 1979).

Use of zinc-efficient genotypes offers a low-input approach for overcoming zinc deficiency and minimizing fertilizer use for optimum yields. This approach is especially effective in management of zinc deficiency when it is induced due to adverse soil chemical conditions, such as calcareousness (Sakal et al. 1988; Cakmak et al. 1997a) and sodicity (Qadar, 2002). Recently, Singh et al. (2005) have discussed the factors contributing to Zn-efficiency of cereals and suggested that this could be improved through manipulation of mechanisms involved in synthesis and release of Zn-phytosiderophores and improving root architecture. The latter has been shown to play an important role in determining the Zn-efficiency of wheat genotypes (Dong et al. 1995).

178

Fig. 11.1. Zn-deficient and Zn-sufficient plants of rice vars. Jagannath, Calrose and Sabarmati (arranged left to right). Grown with Zn-deficient nutrition, var. Jagannath shows severe restricition of growth and Zn-deficiency symptoms, whereas var. Sabarmati shows normal growth, free from Zn deficiency symptoms. Var. Calrose figures intermediate in terms of tolerance to Zn deficiency (Source: Agarwala and Sharma, 1979).

Table 11.7. Some crops investigated for genotypic differences in tolerance to zinc deficiency

Crop	Reference
Wheat (*Triticum aestivum* L.)	Agarwala et al. (1971), Sharma et al. (1971, 1976), Graham et al. (1992), Graham and Rengel (1993), Takkar (1993), Rengel and Graham (1995c), Cakmak et al. (1996c)
Barley (*Hordeum vulgare* L.)	Takkar et al (1983), Gene et al. (2002)
Rice (*Oryza sativa* L.)	Frono et al. (1975), Agarwala et al. (1978), Sakal et al. (1988), Qadar (2002)
Maize (*Zea mays* L.)	Ramani and Kannan (1985)
Triticale (*Secale cereale*)	Cakmak et al. (1997b)
Chickpea (*Cicer arietimum* L.)	Khan et al. (1998)
Soybean (*Glycine max* L.)	Saxena and Chandel (1992)
Navy bean (*Phaseolus vulgaris*)	Jolley and Brown (1991b)
Oilseed rape (*Brassica napus*)	Grewal et al. (1997)
Potato (*Solanum tuberosum* L.)	Sharma and Grewal (1990)

Fig. 11.2. Wheat vars. PV18 (left three pots) and WL 212 (right three pots) raised at 0.0001 μM, 0.001μM and 1.0 μM Zn (arranged left to right). At low (0.0001 and 0.001 μM) Zn supply, var. PV18 shows severe limitation in growth and visible symptoms of Zn deficiency, whereas var. WL 212 shows moderately good growth, and only mild symptoms of Zn-deficiency (Source: Agarwala and Sharma, 1979).

MOLYBDENUM DEFICIENCY

Information on genotypic differences in molybdenum efficiency is limited. In an early study, Young and Takahashi (1953) reported differences in performance of alfalfa cultivars on molybdenum-deficient soils of Hawai and suggested avoiding cultivation of the susceptible cultivars as forage. Based on growth depression and severity of visible symptoms induced in response to molybdenum deprivation, Agarwala and Sharma (1979) reported differences in tolerance of several wheat, rice and maize varieties to molybdenum deficiency under sand culture conditions. Wheat var. Sonalika, was reported tolerant to molybdenum deficiency (Fig. 11.3). Franco and Munns (1981) reported differences in molybdenum accumulation in common bean (*Phaseolus vulgaris* L.) genotypes in acid soils, which are low in available molybdenum. In beans, genotypic differences in molybdenum accumulation corresponded with the differences in their N_2 fixation ability (Brodrick and Giller, 1991).

BORON DEFICIENCY

Several crop genotypes are known to show genotypic differences in boron efficiency (Nable et al. 1990; Bellaloui and Brown, 1998; Rerkasem and Jamjod, 1997, 2004, references therein). Based on differences in seed yield

Fig. 11.3. Mo-sufficient and Mo-deficient plants of wheat vars. Sonalika, and WG 377 (arranged in that order from left to right). Grown with Mo-deficient nutrition, var. Sonalika shows apparently normal growth, whereas var. WG 377 exhibits growth depression and symptoms of Mo-deficiency (Source: Agarwala and Sharma, 1979).

on low boron soils of Chiang Mai (Thailand), Rerkasem and associates (see Rerkasem and Jamjod 1997). have reported genotypic differences in boron efficiency in wheat, barley, green gram, black gram and soybean. Wheat cultivars SW 41 and Fang 60 showed marked difference in boron efficiency. The B-inefficient wheat cultivar SW 41 showed even higher sensitivity to boron deficiency than the dicotyledonous cultivars. Wheat has otherwise been rated as highly tolerant to boron deficiency (Martens and Westermann, 1991). Rerkasem and Jamjod (2004) have suggested that advanced lines of wheat genotypes, found to be B-efficient on the basis of performance (yield) on low boron soils, should be used for development of B-efficient varieties. This has advantage over the use of the international wheat germplasm available with the International Center for Maize and Wheat Research (CIMMYT) in breeding for tolerance to B deficiency as the latter is reported to be largely B-inefficient (Rerkasem and Jamjod, 2004).

REFERENCES

Abadia A, Poc A, Abadia J (1991). Could iron nutrient status be evaluated through photosynthetic pigment changes? *J. Plant Nutr.* **14**: 987-999.

Abadia J (1992). Leaf responses to Fe deficiency: A review. *J. Plant Nutr.* **15**: 1699-1713.

Abadia J (ed.) (1995). *Iron nutrition in Soils and Plants.* Kluwer Academic Publishers, Dordrecht.

Abreu CA, Van Raj B, Gabe U, Abreu MF, Paz-Gonzalez A (2002). Efficiency of multinutrient extractants for the determining of available zinc in soils. *Commun. Soil Sci. Plant Anal.* **33**: 3313-3324.

Adams JF (1997). Field response to molybdenum by field and horticultural crops. In: *Molybdenum in Agriculture,* UC Gupta, (ed.) Cambridge University Press, Cambridge 182-201.

Adams JF, Burmester CH, Mitchel CC (1990). Long-term fertility treatments and molybdenum availability. *Fert. Res.* **21**: 167-170.

Adams ML, Norvell WA, Philpot WD, Peverley JH (2000). Towards the determination of manganese, zinc, copper and iron deficiency in 'Bragg' soybean using spectral detection methods. *Agronomy J.* **92**: 268-274.

Adams P, Graves CJ, Winson GW (1975). Some effects of copper and boron deficiencies on the growth and flowering of *Chrysanthemum morifolium* cv. Hurricane *J. Sci. Food. Agric* **26**: 899-901.

Adriano DC (1986). *Trace Elements in Terrestrial Environment.* Springer-Verlag, New York.

Afridi MMRK, Hewitt EJ (1964). The inducible formation and stability of nitrate reductase in higher plants I. Effects of nitrate and molybdenum on enzyme activity in cauliflower (*Brassica oleracea* var. botrytis). *J. Exp. Bot.* **15**: 251-261.

Agarwala SC, Abidi A, Sharma CP (1991). Variable boron supply and sugarbeet metabolism. *Proc. Natl. Acad. Sci.* (India) **61**(BI): 109-114.

Agarwala SC, Bisht SS, Sharma CP (1977a). Relative effectiveness of certain heavy metals in producing toxicity and iron deficiency in barley. *Can. J. Bot.* **55**: 1299-1307.

Agarwala SC, Bisht SS, Sharma CP, Afzal A (1976). Effect of deficiency of certain micronutrients on the activity of aldolase in radish plants grown in sand culture. *Can. J. Bot.* **54**: 76-78.

Agarwala SC, Chatterjee C, Nautiyal N (1986a). Effect of manganese supply on the physiological availability of iron in rice plants grown in sand culture. *Soil Sci.Plant Nutr.* **32**: 169-178.

Agarwala SC, Chatterjee C, Nautiyal N (1988). Effect of induced molybdenum deficiency on growth and enzyme activity in sorghum. *Trop. Agric.* **65**: 333-336.

Agarwala SC, Chatterjee C, Sharma CP (1978a). Relative susceptibility of some high-yielding rice varieties to Zn deficiency. In. *Environmental Physiological Ecology of Plants.* D. Sen, RP Bansal (eds) Bishen Singh and Mahendra Pal Singh, Dehra Dun. 41-50.

Agarwala SC, Chatterjee C, Sharma PN, Sharma CP, Nautiyal N (1979). Pollen development in maize plants subjected to molybdenum deficiency. *Can. J. Bot.* **57**: 1946-1950.

Agarwala SC, Farooq S, Sharma CP (1977b). Growth and metabolic effects of boron deficiency in some plant species of economic importance. *Geophytology* **7**: 79-90.

Agarwala SC, Hewitt EJ (1954). Molybdenum as a plant nutrient. IV The interrelationships of molybdenum and nitrate supply in chlorophyll and ascorbic acid fractions in cauliflower plants grown in sand culture. *J. Hort. Sci.* **29**: 291-300

Agarwala SC, Hewitt EJ (1955). Molybdenum as a plant nutrient V. The interrelationship of molybdenum and nitrate supply in the concentration of sugars, nitrate and organic nitrogen in cauliflower plants grown in sand culture. *J. Hort. Sci.* **30**: 151-162.

Agarwala SC, Nautiyal BD, Chatterjee C (1986b). Manganese, copper and molybdenum nutrition of papaya. *J. Hort. Sci.* **61**: 397-405.

Agarwala SC, Sharma CP (1961). The relation of iron supply to the iron, chlorophyll and catalase in barley plants grown in sand culture. *Physiol. Plant.* **14**: 275-283.

Agarwala SC, Sharma CP (1979). *Recognizing Micronutrient Disorders of Crop Plants on the Basis of Visible Symptoms and Plant Analysis.* Lucknow University, Lucknow.

Agarwala SC, Sharma CP, Farooq S (1965). Effect of iron supply on growth, chlorophyll, tissue iron and activity of certain enzymes in maize and radish. *Plant Physiol.* **40**: 493-499.

Agarwala SC, Sharma CP, Farooq S, Chatterjee C (1978b). Effect of molybdenum deficiency on the growth and metabolism of corn plants raised in sand culture. *Can. J. Bot.* **56**: 1905-1908.

Agarwala SC, Sharma CP, Kumar A (1964). The interelationship of iron and manganese supply in growth, chlorophyll, and iron porphyrin enzymes in barley plants grown in sand culture. *Plant Physiol.* **39**: 603-609.

Agarwala SC, Sharma CP, Sharma PN, Nautiyal BD (1971). Susceptibility of some high yielding varieties of wheat to deficiency of micronutrients in sand culture. *Proc. Intl. Symp. Soil Fert. Evaln.* New Delhi Vol. 1. JS Kanwar, NP Datta, DR Bhumbla, TD Biswas (eds) Indian Soc. Soil Sci. New Delhi. 1047-1063.

Agarwala SC, Sharma PN, Chatterjee C, Sharma CP (1980). Copper deficiency induced changes in wheat anther. *Proc. Indian Natl. Sci. Acad.* **B46**: 172-176.

Agarwala SC, Sharma PN, Chatterjee C, Sharma CP (1981). Development and enzymatic changes during pollen development in boron deficient maize plants. *J. Plant Nutr.* **3**: 329-336.

Ahmed S, Evans HJ (1960). Cobalt: A micronutrient for the growth of soybean plants under symbiotic conditions. *Soil Sci.* **90**: 205-210.

Aitken RL, Jeffrey AJ, Compton BL (1987). Evaluation of selective extractants for boron in some Queensland soils. *Aust. J. Soil Res.* **25**: 263-273.

Alcantara E, de la Guardia MD (1991). Variability of sunflower inbred lines to iron deficiency stress. *Plant Soil.* **130**: 93-96.

Alcaraz CF, Hellin E, Sevilla, F, Martinez-Sánchez F (1985). Influence of the leaf iron content on the ferredoxin levels in citrus plants. *J. Plant Nutr.* **8**: 603-611.

Alloway BJ, Tills AR (1984). Copper deficiency in world crops. *Outlook Agric.* **13**: 32-42

Alscher RG, Erturk, N, Heath LS (2002). Role of superoxide dismutases (SODs) in controlling oxidative stress. *J. Exp. Bot.* **53**: 1331-1341.

Ananyev GM, Dismukes GC (1997). Calcium induces binding and foRNAtion of a spin coupled dimanganese (Mn II, Mn II) center in the apo-water oxidation complex of photosystem II as precursor to the functional tetra Mn/Ca cluster. *Biochemistry* **36**: 11342-11350.

Andrews SC, Arosio P, Bottke W, Briat JF, Von Darl M, Harrison PM, Laulhere JP, Levi S, Lobreaux S, Yewdall SJ (1992). Structure, function and evolution of ferritins. *J. Inorg. Biochem.* **47**: 161-174.

Apel K, Hirt H (2004). Reactive oxygen species: Metabolism, oxidative stress and signal transduction. *Annu. Rev. Plant Biol.* **55**: 373-400.

Appel HM (1993). Phenolics in ecological interactions. The importance of oxidation. *J. Chem. Ecol.* **9**: 1521-1552.

Appleby CA, Bogusz D, Dennis ES, Peacock WJ (1988). A role for haemoglobin in all plants roots? *Plant Cell Environ.* **11**: 359-367.

Arnon DI (1954). Criteria of essentiality of inorganic micronutrients for plants. In *Trace Elements and Plant Physiology.* Chronica Botanica. T Wallace (Ed.), Waltham, Mass. 31-39.

Arnon DI, Stout PR (1939). The essentiality of certain elements in minute quantity for plants with special reference to copper. *Plant Physiol.* **14**: 371-375.

Arulanathan AR, Rao IM, Terry N (1990). Limiting factors in photosynthesis VI. Regeneration of ribulose-1,5-biphosphate limits photosynthesis at low photochemical capacity. *Plant Physiol.* **93**: 1466-1475.

Asad A, Bell RW, Dell B, Huang L (1997). Development of a boron buffered solution culture system for controlled studies of plant nutrition. *Plant Soil* **188**: 21-32.

Asad A, Blamey FPC, Edwards DG (2003). Effects of boron foliar applications on vegetative and reproductive growth of sunflower. *Ann. Bot.* **92**: 565-570.

Asada K (1992). Ascorbate peroxidase: a hydrogen peroxide scavenging enzyme in plants. *Plant Physiol.* **85**: 235-241.

Asada K (1994). Production and action of active oxygen species in photosynthetic tissues. In: *Causes of Photooxidatve Stress and Amelioration of Defense Systems in Plants.* CH Foyer and PM Mullineaux (eds). CRC Press, Boca Raton, Florida. 77-104.

Asada K (1997). The role of ascorbate peroxidase and mono-dehydroascorbate reductase in H_2O_2 scavenging in plants. In *Oxidation States and the Molecular Biology of Antioxidant Defenses.* JG Scandalios (ed.) Cold Spring Harbor Laboratory Press. 715-735.

Asher CJ (1981). Limiting external concentration of the trace elements for plant growth: Use of flowing solution culture technique. *J. Plant Nutr.* **3**: 163-180.

Asher CJ, Edwards DG (1983). Modern solution culture techniques. In *Encyclopedia of Plant Physiology 15A: Inorganic Plant Nutrition.* A Lauchli, RL Bieleski (eds). Springer Verlag Berlin, Hidelberg, New York, Tokyo. 94-119.

Assunção AGL, Costa Martins PDA, De Folter S, Vooijs R, Schat H, Aarts MGM (2001). Elevated expression of metal transporter genes in three accessions of the metal hyperacumulator *Thlaspi caerulescens*. *Plant Cell Environ.* **24**: 217-226.

Axelsen KB, Palmgren MG (2001). Inventory of the superfamily of P-type ion pumps in *Arabidopsis*. *Plant Physiol.* **126**: 696-706.

Ayala MB, Lopez GJ, Lachica M, Sandmann G (1992). Changes in carotenoids and fatty acids in photosystem II of Cu-deficient pea plants. *Physiol Plant.* **84**: 1-5.

Ayala MB, Sandmann G (1988). Activities of Cu-containing proteins on Cu-depleted pea leaves. *Physiol. Plant* **72**: 801-806.

Azouaou Z, Souvré A (1993). Effects of copper deficiency on pollen fertility and nucleic acids in the durum wheat anther. *Sex. Plant Reprod.* 6: 199-209.

Backer JE, Gauch HG, Duggar Jr. WM (1956). Effects of boron on the water relations of higher plants. *Plant Physiol.* **31**: 89-94.

Badger M, Price DG (1994). The role of carbonic anhydrase in photosynthesis. *Annu. Rev. Plant Physiol. Plant Mol. Biol.* **45**: 369-392.

Bagnaresi P, Basso B, Pupillo P (1997). The NADH-dependent Fe^{3+}-chelate reductase of tomato roots. *Planta* **202**: 427-434.

Baluska F, Hlavacka A, Šamaj J, Palme K, Robnison DG, Matoh T, McCurdy DW, Menzel D, Volkman D (2002). F-actin dependent endocytosis of cell wall pectins in meristematic root cells. Insights from brefeldin A–induced compartments. *Plant Physiol.* **130**: 422-431.

Bandurski RS, Cohen JD, Slovin JP, Reinecke DM (1995). Auxin biosynthesis and metabolism. In *Plant Hormones.* PJ Davis (ed.) Kluwer Academic Publishers, Dordrecht, 39-65.

Bansal RL, Nayyar VK, Takkar PN (1991). Field screening of wheat cultivars for manganese efficiency. *Field Crops Research* **29**: 107-112.

Bar-Akiva A (1984). Substitutes for benzidine as H-donor in the peroxidase assay, for rapid diagnosis of iron deficiency in plants. *Commun. Soil Sci. Plant Anal.* **15**: 929-934.

Bar-Akiva A, Lavon R (1969). Carbonic anhydrase activity as an indicator of zinc deficiency in citrus leaves. *J. Hort. Sci.* **44**: 359-362.

Bar-Akiva A, Lavon R, Sagiv J (1969). Ascorbic acid oxidase activity as a measure of the copper nutrition requirement of citrus leaves. *Agrochim.* **14**: 47-54.

Bar-Akiva A, Lavon R (1967). Visible symptoms of some metabolic patterns in micronutrient deficient Eureka lemon leaves. *Israel J. Agr. Res.* **17**: 7-16.

Bar-Akiva A, Sagiv J, Hasdai D (1971). Effect of mineral deficiencies and other cofactors on the aldolase enzyme activity of citrus leaves. *Physiol. Plant.* **25**: 386-390.

Barber MJ, Kay CJ (1996). Superoxide production during reduction of molecular oxygen by

Barber SA (1995). *Soil Nutrient Bioavailability: A Mechanistic Approach* 2nd ed. John Wiley.

Baron M, Arellano B, Lopez-GJ (1995). Copper and photosystem II. A controvertial relationship. *Physiol. Plant.* **94**: 174-180.

Baron M, Lachica M, Chueca A, Sandmann G (1990). The role of copper in the structural organisation of photosystem II in chloroplast membranes. In *Current Research in Photosynthesis* Vol. I. M. Baltscheffsky (ed.). Kluwer Academic Publishers. Dordrecht. 303-306.

Barr R, Böttger M, Crane FL (1993). The effect of B on plasma membrane electron transport and associated proton secretion by cultured carrot cells. *Biochem. Mol. Biol. Int.* **31**: 31-39.

Barr R, Crane FL (1974). A study of chloroplast membrane polypeptides from mineral deficient maize in relation to photosynthetic activity. *Proc. Indiana Acad. Sci.* **83**: 95-100.

Barr R, Crane FL (1991). Boron stimulates NADH oxidase activity of cultured carrot cells. In *Current Topics in Plant Biochemistry and Physiology.* Vol 10. DD Randall, DG Blevins, CD Miles (eds) Univ. Mo. Press, Columbia. 290

Baszyński T, Ruszkowska M, Król M, Tukendorf A, Wolińska DI (1978). The effect of copper deficiency on the photosynthetic apparatus of higher plants. *Z. Pflanzenphysiol* **89**: 207-216.

Bauma D (1983). Diagnosis of mineral deficiencies using plant tests. In *Encyclopedia of Plant Physiology 15A: Inorganic Plant Nutrition.* A Lauchli, RL Bieleski (eds). Springer Verlag Berlin, Hidelberg, New York, Tokyo. 120-146.

Baylock AD (1995). Navybean yield and maturity response to nitrogen and zinc. *J.Plant Nutr.* **18**: 163-178.

Becana M, Maran JF, Iturbe-Ormaetxe I (1998). Iron dependent oxygen free radical generation in plants subjected to environmental stress: Toxicity and antioxidant protection. *Plant Soil* **201**: 137-147.

Becker R, Grün M, Scholz G (1992). Nicotianamine and the distribution of iron in the apoplasm and symplasm of tomato (*Lycopersicon esculentum* Mill.) *Planta* **187**: 48-52.

Bell PF, Chaney RL, Anglle JS (1991). Determination of the free Cu^{2+} activity required by corn (Zea mays L) using chelator-buffered nutrient solutions. *Soil Sci. Soc. Amer. J.* **55**: 1366-1374.

Bell RW, McLay L, Plaskett D, Dell B, Loneragan JF (1989). Germination and vigour of black gram (*Vigna mungo* (L.) Hepper) seed from plants grown with and without boron. *Aust. J. Agric. Res.* **40**: 273-279.

Bellaloui N, Brown PH (1998). Cultivar differences in boron uptake and distribution in celery (*Apium graveolens*), tomato (*Lycopersicon esculentum*) and wheat (*Triticum aestivum*). *Plant Soil* **198**: 153-158.

Bellaloui N, Brown PH, Dandekar AM (1999). Manipulation of *in vivo* sorbitol production alters boron uptake and transport in tobacco. *Plant Physiol.* **119**: 735-741.

Belouchi A, Kwan T, Gros P (1997). Cloning and characterization of the *Os nramp* family from *Oryza sativa*, a new family of membrane proteins possibly implicated in the transport of metal ions. *Plant Mol. Biol.* **33**: 1085-1092.

Bennett WR (ed.) (1993). *Nutrient Deficiencies & Toxicities in Crop Plants.* APS Press. St Paul Minn.

Bereczky Z, Wang H-Y, Schubert V, Ganal M, Bauer P (2003). Differential regulation of *nramp* and *irt* metal transporter genes in wild type and iron uptake mutants of tomato. *J. Biol. Chem.* **278**: 24697-24704.

Berg JM Tymoczko JL, Stryer L (2002). *Biochemistry.* 5th ed. WH Freeman and Company, New York.

Berg JM, Shi Y (1996). The galvanization of biology: a growing appreciation for the role of zinc. *Science.* **271**: 1081-1085.

Berger KC, Truog E (1939). Boron determination of soils and plants *Ind. Eng. Chem. Anal. Ed.* **11**: 540-545.

Bergmann W (1992). *Nutritional Disorders of Plants- Development, Visual and Analytical Diagnosis.* Fischer Verlag, Jena.

Berry JA, Reisenauer HM (1967). The influence of molybdenum on iron nutrition of tomato.

Bertamini M, Muthuchelian K, Nedunchezhian N (2002). Iron deficiency induced changes on the donor side of PS II in field grown grapevine (*Vitis vinefera* L. cv. Pinot noir) leaves. *Plant Science* **162**: 599-605.

Berthold DA, Andersson ME, Nordlund P (2000). New insight into the structure and function of the alternative oxidase. *Biochem. Biophys. Acta.* **1460**: 241-254.

Berthold DA, Stenmark P (2003). Membrane-bound diiron carboxylate proteins. *Annu. Rev. Plant Biol.* **54**: 497-517.

Bettger WJ, O'Dell BLA (1981). Critical physiological role of Zn in structure and function of biomembranes. *Life Sci.* **28**: 1425-1438.

Bienfait HF (1985). Regulated redox processes at the plasmalemma of plant root cells and their function in iron uptake. *J. Bioenerg. Biomember.* **17**: 73-83.

Bienfait HF (1988). Mechanisms in Fe-efficiency reactions of higher plants. *J. Plant Nutr.* **11**: 605-629.

Bienfait HF, Bino RJ, Van der Blick AM, Duivenvoorden JF, Fontain IM (1983). Characterization of ferric reducing activity in roots of Fe-deficient *Phaseolus vulgaris. Physiol. Plant.* **59**: 196-202.

Bienfait HF, De Weger LA, Kramer D (1987). Control of the development of iron-efficiency reactions in potato as a response to iron deficiency is located in the roots. *Plant Physiol.* **83**: 244-247.

Bigham JM (ed) (1996). *Methods of Soil Analysis.* Soil Sci. Soc. Amer. Madison WI.

Bindra A (1980). Iron chlorosis in horticultural and field crops. *Annu. Rev. Plant Sci.* **2**: 221-312.

Bingham, FT (1982). Boron. In *Methods of Soil Analysis* Part 2. Chemical and Microbiological Properties. AL Page, RH Miller, D Keeney (eds) Amer. Soc. Agron. & Soil Sci. Soc. Amer. Madison WI. 431-446.

Birnbaum EH, Beasley CA, Dugger WM (1974). Boron deficiency in unfertilized cotton (*Gossypium hirsutum*) ovules grown in vitro. *Plant Physiol.* **54**: 931-935.

Bisht SS, Nautiyal BD, Sharma CP (2002a) Biochemical changes under iron deficiency and recovery in tomato. *Indian J. Plant Physiol.* **7**: 183-186.

Bisht SS, Nautiyal BD, Sharma CP (2002b). Zinc nutrition dependent changes in tomato (*Lycopersicon esculentum* Mill) metabolism. *J. Plant Biol.* **29**: 159-163.

Black CA (1993). *Soil Fertility Evaluation and Control.* Lewis Publishers, Boca Raton Fl.

Blaser-Grill J, Knoppik D, Amberger A, Goldbach H (1989). Influence of boron on the membrane potential of *Elodea densa* and *Helianthus annuus* roots and H^+ extrusion of suspension cultured *Daucus carota* cells. *Plant Physiol.* **90**: 280-284.

Blevins DG, Lukaszewski KM (1998). Boron in plant structure and function. *Ann. Rev. Plant Physiol. Plant Mol. Biol.* **49**: 481-501.

Bligny R, Douce R (1977). Mitochondria of isolated plant cells (*Acer pseudoplatanus* L.) II, Copper deficiency effects on cytochrome c oxidase and oxygen uptake. *Plant Physiol.* **60**: 675-679.

Boawn LC, Rasmussen PE, Brown JE (1969). Relationship between tissue zinc levels and maturity period of field beans. *Agron.J.* **61**: 49-51.

Bohnsack CW, Albert LS (1977). Early effects of boron deficiency on indole acetic acid oxidase levels of squash root tips. *Plant Physiol.* **59**: 1047-1050.

Bolanos L, Brewin NJ, Bonilla I (1996). Effects of boron on *Rhizobium*-legume cell-surface interactions and nodule development. *Plant Physiol.* **110**: 1249-1256.

Bonilla I, Cadahia C, Carpena O, Hernando V (1980). Effect of boron on nitrogen metabolism and sugar levels of sugarbeet. *Plant Soil* **57**: 3-9.

Bonilla I, Garcia-GonŸalez M, Mate O (1990). Boron requirement in cyanobacteria. Its possible role in the early evolution of photosynthetic organisms. *Plant Physiol.* **94**: 1554-1560.

Bonilla I, Mergold-Villaseñsor C, Campos ME, Sánches N, Pérez H, Lopez L, Castrejon L, Sanchez F, Cassab GI (1997). The aberrant cell walls of boron-deficient bean root nodules have no covalantly bound hydroxyproline-proline-rich proteins. *Plant Physiol.* **115**: 1329-1340.

Bonnet M, Camares O, Veisseire P (2000). Effect of zinc and influence of *Acremonium lolii* on growth parameters, chlorophyll *a* fluorescence and antioxidant enzyme activities of ryegrass (*Lolium perenne* L. cv Apollo). *J. Exp. Bot.* 51: 945-953.

Botrill DE, Passingham JV, Kriedmann PE (1970). The effect of nutrient deficiencies on photosynthesis and respiration in spinach. *Plant Soil* **32**: 424-438.

Bould C, Hewitt EJ, Needham P (1983). *Diagnosis of Mineral Disorders of Plants.* Vol. 1.

Bouzayen M, Felix G, Latché A, Pech J-C, Boller T (1991). Iron: An essential cofactor for the conversion of 1-aminocyclopropane-1-carboxylic acid to ethylene. *Planta* **184**: 244-247.

Bouzayen M, Latché A, Pech JC (1990) sub-cellular localization of the sites of conversion of 1-aminocyclopropane-1-carboxylic acid into ethylene in plant cells. *Planta* **180**: 175-180.

Bove JM, Bove C, Whatley FR, Arnon DI (1963). Chloride requirement for oxygen evolution in photosynthesis. *Z. Naturforsch.* **18b**: 683-688.

Bowler C, Slooten L, Vanderbnanden S, De Rycke R, Botterman J, Sybesma C, Van Montagu M, Inze D (1991). Manganese superoxide dismutase can reduce cellular damage mediated by oxygen radicals in transgenic plants. *EMBO J* **10**: 1723-1732.

Bozak KR, Yu H, Sirevåg R, Chistoffersen RE (1990). Sequence analysis of ripening-related cytochrome P-450 cDNAs from avocado fruit. *Proc. Natl. Acad. Sci. USA* **87**: 3904-3908.

Brady NC, Weil RR (ed) (1999). *The Nature and Properties of Soils.* 12th ed. Prentice Hall Inc. International (UK), London.

Brenchly WE, Thornton BA (1925). The relation between development, structure and functioning of the nodules on *Vicia faba*, as influenced by the presence or absence of boron in the nutrient medium. *Proc. R. Soc. London.* Ser. B. Biol. Sci. **98**: 373-398.

Brennan RF (1992). The role of manganese and nitrogen nutrition in the susceptibility of wheat plants to take-all in Western Australia. *Fert. Res.* **31**: 35-41.

Brive A, Sengupta A K, Beuchie D, Larsson J, Kennison JA, Rasmuson-Lestander A, Muller J (2001). *Su (Z) 12,* a novel *Drosophila* Polycomb-group gene that is conserved in vertebrates and plants. *Development.* **128**: 3371-3379.

Brodrick SJ, Giller KE (1991). Genotypic difference in molybdenum accumulation affects N_2-fixation in tropical *Phaseolus vulgaris. J. Exp. Bot.* **42**: 1339-1343.

Brodrick SJ, Sakala MK, Giller KE (1992). Molybdenum reserves of seed, and growth and N_2 fixation by *Phaseolus vulgaris* L. *Biol. Fert. Soils* **13**: 39-44.

Browman MG, Chestess G, Pionke HB (1969). Evolution of tests for predicting the availability of manganese to plants. *J. Agri. Sci.* **72**: 335-340.

Brown JC (1961). Iron chlorosis in plants. *Adv. Agron.* **13**: 329-369.

Brown JC (1963). Iron chlorosis in soybeans as related to the genotype of rootstock. *Soil Sci.* **98**: 387-394.

Brown JC (1978). Mechanism of iron uptake by plants. *Plant Cell Environ.* **1**: 249-257.

Brown JC (1979). Effects of boron stress on copper enzyme activities in tomato. *J. Plant Nutr.* **1**: 39-53.

Brown JC, Ambler JE (1973). Reductants released by roots of Fe-deficient soybean. *Agron. J.* **65**: 311-314.

Brown JC, Ambler JE, Chaney RL, Foy CD (1972). Differential responses of plant genotypes to micronutrients: In *Micronutrients in Agriculture*. JJ Mortvedt, PM Giordano, WL Lindsay (eds) Soil Sci Soc. Amer. Madison WI. pp 389-418.

Brown JC, Chaney RL, Ambler JE (1971). A new tomato mutant inefficient in the transport of iron. *Physiol. Plant.* **25**: 45-58.

Brown JC, Clark RB (1977). Copper as essential to wheat reproduction. *Plant Soil.* **48**: 509-523.

Brown JC, Hendricks, SB (1952). Enzymatic activities as indication of copper and iron deficiencies in plants. *Plant Physiol.* **27**: 651-660.

Brown JC, Jolley VD (1989). Plant metabolic responses to iron deficiency stress. *Bioscience.* **39**: 546-551.

Brown JC, Jones WE (1971). Differential transport of boron in tomato. *Plant Physiol.* **25**: 279-282.

Brown PH, Bellaloui N, Hu H, Dandekar A (1999). Transgenically enhanced sorbitol synthesis facilitates phloem boron transport and increases tolerance of tobacco to boron deficiency. *Plant Physiol.* **119**: 17-20.

Brown PH, Bellaloui N, Wimmer MA, Bassil ES, Ruiz J, Hu H, Pfeffer H, Dannel F, Römheld V (2002). Boron in plant biology. *Plant biol.* **4**: 205-223.

Brown PH, Graham RD, Nicholas DJD (1984). The effects of manganese and nitrate supply on the level of phenolics and lignin in young wheat plants. *Plant Soil* 81: 437-440.

Brown PH, Hu H (1994). Boron uptake by sunflower, squash and cultured tobacco cells. *Physiol Plant.* **91**: 435-441.

Brown PH, Shelp BJ (1997). Boron mobility in plants. *Plant Soil* **193**: 85-101.

Brown PH, Welch RM, Carey EE (1987). Nickel: An essential micronutrient for higher plants. *Plant Physiol.* **85**: 801-803.

Broyer TC, Carlton AB, Johnson CM, Stout PR (1954). Chlorine—A micronutrient element for higher plants. *Plant Physiol.* **29**: 526-532.

Brüggemann W, Moog PR, Nakagawa H, Janiesch P, Kupier PJC (1990). Plasma membrane- bound NADH:Fe^{3+}-EDTA reductase and iron deficiency in tomato (*Lycopersicon esculentum*). Is there a Turboreductase? *Physiol. Plant.* **79**: 339-346.

Bughio N, Yamaguchi H, Nishizawa NK, Nakanishi H, Mori S (2002). Cloning an iron regulated metal transporter from rice. *J. Exp. Bot.* **53**: 1677-1682.

Bukovac JM, Wittiver SH (1957). Absorption and mobility of applied nutrients. *Plant Physiol.* **32**: 428-435.

Burnell JN (1986) Purification and properties of phosphoenol pyruvate carboxykinase from C_4 plants. *Aust. J. Plant Physiol.* **13**: 577-587.

Burnell JN (1988). The biochemistry of manganese in plants. In *Manganese in Soils and Plants*. RD Graham, RJ Hannam, NC Uren (eds). Kluwer Academic, Dordrecht. 125-137.

Burnell JN, Hatch MD (1988). Low bundle sheath carbonic anhydrase is apparently essential for effective C_4 pathway operation *Plant Physiol.* **86**: 1252-1256.

Bussler W (1981a). Microscopic possibilities for the diagnosis of trace element stress in plants. *J. Plant Nutr.* **3**: 115-128.

Bussler W (1981b). Physiological functions and utilization of copper. In *Copper in Soils and Plants.* J.F. Loneragan, AD Robson, RD Graham (eds). Academic Press, London. 213-234.

Caballero R, Arouzo M, Hernaiz PJ (1996). Accumulation and redistribution of mineral elements in common vetch during pod filling. *Agron J.* **88**: 801-805.

Cairns ALP, Kurtzinger JH (1992). The effect of molybdenum on seed dormancy in wheat. *Plant Soil.* **145**: 295-297.

Cakamk I, Marschner H (1988a). Increase in membrane permeability and exudation in roots of cotton deficient plants. *J. Plant Physiol.* **132**: 356-361.

Cakamk I, Marschner H (1988b). Zinc-dependent changes in ESR signals, NADPH-oxidase and plasma membrane permeability in cotton roots. *Physiol. Plant* **73**: 182-186.

Cakmak I (2000). Possible roles of zinc in protecting plant cells from damage by reactive oxygen spices. *New Phytol.* **146**: 185-205.

Cakmak I, Alti M, Kaya R, Evliya H, Marschner H (1997a). Differential response of rye, bread and durum wheat to zinc deficiency in clacareous soils. *Plant Soil* **188**: 1-10.

Cakmak I, Derici R, Torun B, Tolay I, Braun HJ, Schlegel (1997b). Role of rye chromosome in improvement of zinc efficiency in wheat and triticale. *Plant Soil* **196**: 249-253.

Cakmak I, Engels C (1999). Role of mineral nutrients in photosynthesis and yield formation. In *Mineral Nutrition of Crops: Fundamental Mechanisms and Implications.* Z Rengel (ed.) Haworth Press. New York, USA: 141-168.

Cakmak I, Erenoglu B, Gülüt KY, Derici R, Römheld V (1998). Light mediated release of phytosiderophores in wheat and barley under Fe or Zn deficiency. *Plant Soil* **202**: 309-315.

Cakmak I, Gülüt KY, Marschner H, Graham RD (1994). Effect of zinc and iron deficiency on phytosiderophore release in wheat genotypes differing in zinc efficiency. *J. Plant Nutr.* **17**: 1-17.

Cakmak I, Kurz H, Marschner H (1995). Short-term effects of boron, germanium and high light intensity on membrane permeability in boron-deficient leaves of sunflower. *Physiol. Plant.* **95**: 11-18.

Cakmak I, Marschner H (1986). Mechanism of phosphorus-induced zinc deficiency in cotton. I. Zinc deficiency-induced uptake rate and phosphorus. *Physiol. Plant.* **68**: 483-490.

Cakmak I, Marschner H (1987). Mechanism of phosphorus-induced zinc deficiency in cotton III changes in physiological availability of zinc in plants. *Physiol. Plant.* **70**: 13-20.

Cakmak I, Marschner H (1988c). Enhanced superoxide radical production in roots of zinc deficient plants. *J. Exp. Bot.* **39**: 1449-1460.

Cakmak I, Marschner H (1993). Effect of zinc nutritional status on activities of superoxide radical and hydrogen peroxide scavenging enzymes in bean leaves. *Plant Soil* **155/156**: 127-130.

Cakmak I, Marschner H, Bangerth F (1989). Effect of zinc nutritional status on growth, protein metabolism and levels of indole-3-acetic acid and other phytohormones in bean (*Phaseolus vulgaris* L.). *J. Exp. Bot.* **40**: 405-412.

Cakmak I, Ozturk L and Karanlibe S (1996a). Zinc efficient wild grasses enhanced release of phytosideraphores under zinc deficiency. *J. Plant Nutr.* **19**: 551-563.

Cakmak I, Oztürk L, Eker S, Torun B, Kalfa HI, Yilmaz A (1997c). Concentration of zinc and activity of copper/zinc-superoxide dismutase in leaves of rye and wheat cultivars differing in sensitivity to zinc deficiency. *J. Plant Physiol.* **151**: 91-95.

Cakmak I, Römheld V (1997). Boron deficiency induced impairment of cellular functions in plant. *Plant Soil.* **193**: 71-83.

Cakmak I, Sari N, Marschner H, Ekiz H, Kalayci M, Yilmaz A, Braun HJ (1996b). Phytosiderophse release in bread and durum wheat genotypes differing in zinc efficiency. *Plant Soil* **180**: 183-189.

Cakmak I, Sari N, Marschner H, Kalayci M, Yilmaz A, Eker S, Gulut KY (1996c). Dry matter production and distribution of zinc in bread and durum wheat genotypes differing in zinc efficiency. *Plant Soil* **180**: 173-181.

Cakmak I, Yilmaz A, Kalayci M, Ekiz H, Torum B, Erenoðlu B, Braun HJ (1996d). Zinc deficiency as a critical problem in wheat production in Central Anatolia. *Plant Soil* **180**: 165-172.

Camacho-Christobal JJ, Gonzalez-Fontes A (1993). Boron deficiency causes drastic decrease in nitrate reductase activity and increases the content of carbohydrates in leaves from tobacco plants. *Planta.* **209**: 528-536.

Camacho-Cristóbal JJ, Anzellotti D, González-Fontes A (2002). Changes in phenolic metabolism of tobacco plants during short- term boron deficiency. *Plant Physiol. Biochem.* **40**: 997-1002.

Camacho-Cristobal JJ, Lunar L, Lafont F, Baumert A, Gonzélez-Fontes A (2004). Boron deficiency causes accumulation of chlorogenic acid and caffeoyl polyamine conjugates in tobacco leaves. *J. Plant Physiol.* **161**: 879-881.

Camacho-Cristóbal JJ, Maldonado JM, González–Fontes A (2005). Boron deficiency increases putrescine levels in tobacco plants. *J. Plant Physiol.* **162**: 921-928.

Camp SD, Jolley VD, Brown JC (1987). Comparative evaluation of factors involved in Fe stress response in tomato and soybean. *J. Plant Nutr.* **10**: 423-442.

Campbell KA, Force DA, Nixon PJ, Dole F, Diner DB, Britt RD (2000). Dual mode EPR detects initial intermediate in photoassembly of photosystem II Mn cluster: Influence of the aminoacid residue 170 on the D1 polypeptide on manganese coordination. *J. Amer. Chem. Soc.* **122**: 3754-3761.

Campbell LC, Nable RO (1988). Physiological functions of manganese in plants In *Manganese in Soils and Plants*. RD Graham, RJ Hannan, NC Uren (eds) Kluwer Academic Publishers, Dordrecht. 139-154.

Campbell WH (1999). Nitrate reductase structure, function and regulation: bridging the gap between biochemistry and physiology. *Annu. Rev. Plant Physiol. Plant Mol. Biol.* **50**: 277-303.

Campbell WH (2001). Structure and function of eukaryotic NAD(P)H: nitrate reductase. *Cell Mol. Life Sci.* **58**: 194-204.

Campbell WH, Kinghorn JR (1990). Functional domains of assimilating nitrate reductase and nitrite reductases. *Trends Biochem. Sci.* **15**: 315-319.

Cara FA, Sánchez E, Ruiz JM, Romero L (2002). Is phenol oxidation responsible for the short-term effects of boron deficiency on plasma membrane permeability and function in squash roots? *Plant Physiol. Biochem.* **40**: 853-858.

Carpena RO, Carpena RR, Zoronoza P,Collado G (1984). A possible role of boron in higher plants. *J. Plant Nutr.* **9**: 1341-1354.

Cartwright B, Tiller KG, Zarcinas BA, Spouncer LR (1983). The chemical assessment of the boron status of soils. *Aust. J. Soil Res.* **21**: 321-332.

Caruso C, Chilosi G, Leonardi L, Birtini L, Magro P, Buonocore V, Caporale C (2001). A basic peroxidase from wheat kernel with antifungal activity. *Phytochemistry.* **58**: 743-750.

Casimiro A (1987). Effect of copper on water relations and growth of *Triticum*. In *Plant Response to Stress*. NATO/ASI Series Vol. 15. JD Tenhunen, FM Catarino, OL Lange, WE Oechel (eds) Springer-Verlag, Berlin 459-465.

Casimiro A, Barroso J, Pais MS (1990). The effect of copper deficiency on photosynthetic electron transport in wheat plants. *Physiol. Plant* **79**: 459-464.

Cataldo DA, McFadden KM, Garland TR, Wildung RE (1988). Organic constituents and complexation of nickel (II), iron (III), cadmium (II), and plutonium (IV) in soybean xylem exudates. *Plant Physiol.* **86**: 734-739.

Chandler FB, Miller MC (1946). Effect of boron on the vitamin C content of rutabagas. *Proc. Am. Soc. Hort. Sci.* **47**: 331-334.

Chaney RL (1984). Diagnostic practices to identify iron deficiency in higher plants. *J. Plant Nutr.* **7**: 47-67.

Chaney RL, Bell PF, Coulombe BA(1989). Screening strategies for improved nutrient uptake and use by plants. *Hort. Science.* **24**: 565-572.

Chaney RL, Brown JC, Tiffin LO (1972). Obligatory reduction of ferric chelates in iron uptake by soybeans. *Plant Physiol.* **50**: 208-213.

Chaney RL, Columbe BA, Bell PF, Angle JS (1992b). Detailed method to screen dicot cultivars for resistance to Fe-chlorosis using Fe DTPA and bicarbonate in nutrient solution. *J. Plant Nutr.* **15**: 2063-2083.

Chaney RL, Hamze MH, Bell PF (1992a). Screening chickpea for iron chlorosis resistance using bicarbonate in nutrient solution to simulate calcareous soils. *J. Plant Nutr.* **15**: 2045-2062.

Chapman HD (ed.) (1966). *Diagnostic Criteria for Plants and Soils.* Div. of Agric. Sci. Univ. California, Riverside.

Chapman KSA, Jackson JF (1974). Increased RNA labelling in boron deficient root tip segments. *Phytochemistry* **13**: 1311-1318.

Chapple C (1998). Molecular genetic analysis of plant cytochrome P450-dependent monooxygenases. *Annu. Rev. Plant Physiol. Plant Mol. Biol.* **49**: 311-343.

Chatterjee C, Agarwala SC (1979). Effect of molybdenum and iron supply on molybdenum (^{99}Mo) and iron (^{59}Fe) uptake and activity of certain enzymes in tomato plants grown in sand culture. *J. Nuclear Agric. Biol.* **8**: 21-23.

Chatterjee C, Jain R, Dube BK, Nautiyal N (1998). Use of carbonic anhydrase for determining zinc status of sugarcane. *Tropical Agric.* **75**: 1-4.

Chatterjee C, Nautiyal N (2000). Developmental aberrations in seeds of boron deficient sunflower and recovery. *J. Plant Nutr.* **23**: 835-841.

Chatterjee C, Nautiyal N (2001). Molybdenum stress affects viability and vigour of wheat seeds. *J. Plant Nutr.* **24**: 1377-1386.

Chatterjee C, Nautiyal N, Agarwala SC (1985). Metabolic changes in mustard plants associated with molybdenum deficiency. *New Phytol.* **100**: 511-518.

Chatterjee C, Nautiyal N, Agarwala SC (1994). Influence of changes in manganese and magnesium supply on some aspects of wheat physiology. *Soil Sci. Plant Nutr.* **40**: 191-197.

Chatterjee C, Sinha P, Nautiyal N, Agarwala SC (1989). Physiological response of cauliflower in relation to manganese-boron interaction. *J. Hort. Sci.* **64**: 591-596.

Chaudhry FM, Loneragan JF (1972). Zinc absorption by wheat seedlings and the nature of its inhibition by alkaline earth cations. *J. Exp. Bot.* **23**: 552-560.

Chen Y, Barak P (1982). Iron nutrition of plants in calcareous soils. *Adv. Agron.* **35**: 217-240.

Chen Z-H, Walker RP, Acheson RM, Leegood RC (2002b). Phospho*enol*pyruvate carboxykinase assayed at physiological concentrations of metal ions has a high affinity for CO_2. *Plant Physiol.* **128**: 160-164.

Cheng C, Rerkasem B (1993). Effect of boron on pollen viability in wheat. *Plant Soil.* **155/156**: 313-315.

Cheniae GM, Martin JF (1970). Sites of function of manganese within photosystem II. Roles in O_2 evolution and system II. *Biochem. Biophys. Acta.* **197**: 219-622.

Chereskin BM, Castelfraneo PA (1982). Effect of iron and oxygen on chlorophyll biosynthetic pathway in isolated etio-chlorplast. *Plant Physiol.* **68**: 112-116.

Christensen NW, Powelson RL, Brett (1987). Epidemiology of wheat take-all as influenced by soil pH and temporal changes in inorganic Soil N. *Plant Soil.* **98**: 221-230.

Cianzio S (1999). Breeding crops for improved nutrient efficiency: Soybean and wheat as case studies. In *Mineral Nutrition of Crops: Fundamental Mechanisms and Implications.* Z Rengel (ed.) Haworth Press, Binghamton NY. 267-287.

Cohen CK, Fox TC, Garvin DF, Kochian LV (1998). The role of iron deficiency stress responses in stimulating heavy metal transport in plants. *Plant Physiol.* **116**: 1063-1072.

Cohen CK, Norvell WA, Kochian LV (1997). Induction of the root cell plasma membrane ferric reductase: An exclusive role for Fe and Cu. *Plant Physiol.* **114**: 1061-1069.

Coke L, Whittington WJ (1968). The role of boron in plant growths,IV. Interelationships between boron and indole-3-yl acetic acid in the metabolism of bean radicles. *J. Exp. Bot.* **19**: 295-308.

Coleman JE (1992). Zinc proteins, enzymes, storage proteins, transcription factors, and replication proteins. *Annu. Rev. Biochem.* **61**: 897-946.

Connolly EL, Fett JP, Guerinot ML (2002). Expression of to IRT1 metal transporter is controlled by metals at the levels of transcript and protein accumulation. *Plant Cell* **14**: 1347-1357.

Cooper TG, Filmer D, Wishnick M, Lane MD (1969). The active species of 'CO_2' utilized by ribulose diphosphate carboxylase *J. Biol. Chem.* **244**: 1081-1083.

Cory S, Finch LR (1967). Further studies on the incorporation of phosphate into nucleic acids of normal and boron deficient tissues. *Phytochemistry* **6**: 211-212.

Coulombe BA, Chancey RL, Wiebold WJ (1984a). Use of bicarbonate in screening soybeans for resistance to iron chlorosis. *J. Plant Nutr.* **7**: 411-425.

Coulombe BA, Chaney RL, Wiebold WJ (1984b). Bicarbonate directly induces Fe-chlorosis in susceptible soybean cultures. *Soil Sci. Soc. Am. J.* **48**: 1297-1301.

Cowles JR, Evans HJ, Russel S (1969). B_{12} Co-enzyme dependent ribonucleotide reductase in *Rhizobium* species and the effect of cobalt deficiency on the activity of the enzyme. *J. Bacteriol.* **97**: 1460.

Cox FR (1987). Micronutrient soil tests: Correlation and calibration soil testing: Sampling correlation and interpretation. *Soil Sci. Soc. Amer.* Spl. Publ. No. 21 Madison, WI.: 97-117.

Cox FR, Kamprath EJ (1972). Micronutrient soil tests. In *Micronutrients in Agriculture.* JJ Mortvedt, PM Giordano, WL Lindsay (eds) *Soil Sci. Soc. Amer.* Madison, WI : 289-317

Cram WJ (1983). Chloride accumulation as a homeostatic system: Set points and perturbations. The physiological significance of influx isotherms, temperature effects and the influence of plant growth substances. *J. Exp. Bot.* **34**: 1484-1502.

Crosbie J, Longnecker N, Davies F, Robson A (1993). Effects of seed manganese concentration on lupin emergence. In *Plant Nutrition — From Genetic Engineering to Field Practice* (NJ Barrow, ed.) Kluwer Academic Publishers, Dordrecht. pp. 665-668.

Croteau R, Purkett PK (1989). GPP – synthase. Characterization of the enzyme and evidence that this chain length specific prenyl transferase is associated with monoterpene biosynthesis in sage (*Salvia officinalis*). *Arch. Biochem. Biophys.* **272**: 525-535.

Crozier A, Kamiya Y, Bishop G, Yokota T (2000). Biosynthesis of hormones and elicitor molecules. In *Biochemistry and Molecular Biology of Plants.* B Buchanan, W Gruissem, R Jones (eds) Amer. Soc. Plants Physiol. Rockwille MD 850-929.

Curic C, Alonso JM, Le Jean M, Ecker JR, Briat J-F (2000). Involvement of NRAMP1 from *Arabidopsis thaliana* in iron transport. *Biochem. J.* **347**: 749-755.

Curic C, Briat J-F (2003). Iron transport and signaling in plants *Annu. Rev. Plant Biol.* **54**: 183-206.

Curic C, Panaviene Z, Loulergue C, Dellaporta SL, Briat J-F, Walker EL (2001). Maize *yellow stripe1* encodes a membrane protein involved in Fe(III) uptake. *Nature* **409**: 346-349.

Dai S, Schwendtmayer C, Schriimann P, Ramaswamy S, Eklund H (2000). Redox signaling in chloroplasts: Cleavage of disulphide by an iron–sulphur cluster. *Science.* **287**: 655-658.

Dannel F, Pfeffer H, Romheld V (1997). Effect of pH and boron concentration in the nutrient solution on translocation of boron in the xylem of sunflower. In: *Boron in Soils and Plants.* R W Bell, B Rerkarem (eds) Kluwer Academic Publishers, Dordrecht, The Netherlands; 183-186.

Dannel F, Pfeffer H, Romheld V (1998). Compartmentation of boron in roots and leaves of sunflower as affected by boron supply. *J. Plant Physiol.* **153**: 615-622.

Dannel F, Pfeffer H, Romheld V (2000). Characterization of root boron pools, boron uptake and boron translocation in sunflower using stable isotopes ^{10}B and ^{11}B. *Aust. J. Plant. Physiol.* **27**: 397-405.

Dannel F, Pfeffer H, Römheld V (2002). Update on boron in higher plants-uptake, primary translocation and compartmentation. *Plant biol.* **4:** 193-204.

Dave IC, Kannan S (1980). Boron deficiency and its associated enhancement of RNAse activity in bean plants. *Z. Pflanzenphysiol.* **97:** 261-264.

Davies JN, Adams P, Winsor GW (1978). Bud development and flowering of *Chrysanthemum morifolium* in relation to some enzyme activities and to the copper, iron and manganese status. *Commun. Soil Sci. Plant Anal.* **9**: 249-264.

De Nisi P, Zocchi G (2000). Phosphoenolpyruvate carboxylase in cucumber (*Cucumis sativus* L.) roots under iron deficiency : activity and kinetic characterization. *J. Exptl. Bot.* **51**: 1903-1909.

De Wet E, Robbertse PJ, Groeneveld HT (1989). The influence of temperature and boron on pollen germination in *Mangifera Indica* L. *S. Afr. Tydskr, Plant Grond* **6**: 228-234.

Dean JV, Harper JE (1988). The conversion of nitrite to nitrogen oxide by the constitutive NAD(P)H-nitrate reductase enzyme from soyabean. *Plant Physiol.* **88**: 389-395.

Debus RJ (1992). The manganese and calcium ions in photosynthetic O_2 evolution. *Biochim. Biophys. Acta.* **1102**: 269-352.

Dekock PC, Commisiong K, Farmer VC, Inkson RHE (1960). Interrelationship of catalase, peroxidase, hematin and chlorophyll. *Plant Physiol.* **35**: 599-604.

Del Rio LA, Gómez M, Yañez J, Leal A, López GJ (1978a). Iron deficiency in pea plants. Effect on catalase, peroxidase, chlorophyll and proteins of leaves. *Plant Soil.* **49**: 343-353.

Del Rio LA, Sevitha F, Gomez M, Yanez J, Lopez J (1978b). Superoxide dismutase: An enzyme system for the study of micronutrient interactions in plants. *Planta.* **140**: 221-225.

Delhaize E, Dilworth MJ, Webb J (1986). The effect of copper nutrition and developmental stage on the biosynthesis of diamine oxidase in clover leaves. *Plant Physiol.* **82**: 1126-1131.

Delhaize E, Kataoka T, Hebb DM, White RG, Ryan RR (2003). Genes encoding proteins of the cation diffusion facilitator family that confer manganese tolerance. *Plant Cell* **15**: 1131-1142.

Delhaize E, Loneragan JF, Webb J (1985). Development of three copper metelloenzymes in clover leaves. *Plant Physiol.* **78**: 4-7.

Delhaize E, Loneragan JF,Webb J (1982). Enzymic diagnosis of copper deficiency in subterranian clover II. A simple field test. *Aust. J. Agric. Res.* **33**: 981-987.

Dell B (1981). Male sterility and anther wall structure in copper-deficient plants. *Ann. Bot.* (London) **48**: 599-608.

Dell B, Huang L (1997). Physiological responses of plants to low boron. *Plant Soil.* **193**: 103-120.

Dembitsky V M, Smoum R, Al-Qantar AA, Ali HA, Pergament I, Srebnick M (2002). Natural occurance of boron-containing compounds in plants, algae and microorganisms. *Plant Sci.* **163**: 931-942.

Dembitsky VM, Smoum R, Al-Quntar AA, Ali HA, Pergament I, Srebnik, M (2002). Natural occurence of boron-containing compounds in plants, algae and microorganisms. *Plant Sci.* **163**: 931-942.

Desikan R, Griffiths R, Hancock J, Neill S (2002). A new role for an old enzyme: Nitrate reductase-mediated nitric oxide generation is required for abscisic acid-induced stomatal closure in *Arabidopsis thaliana. Proc. Nat. Acad. Sci.* USA **99**: 16314-16318.

Dhakre G, Singh J, Chauhan SVS (1994). Analysis of phenolics and boron in the stigma of seedless *Campsis grandiflora* K. Schum. *Indian J. Exp. Biol.* **32**: 816-818.

Dilworth MJ, Robson AD, Chatel DL (1979). Cobalt and nitrogen fixation in *Lupinus angustifolius* L. II. Nodule formation and function. *New Phytol.* **83**: 63-79.

Dinkins R, Pflipsen C, Thompson A Collins GB (2002). Ectopic expression of an *Arabadopsis* single zinc finger gene in tobacco results in dwarf plants. *Plant Cell Physiol.* **43**: 743-750.

Dixit D, Srivastava NK, Sharma S (2002). Boron deficiency induced changes in translocation of $^{14}CO_2$-photosynthate into primary metabolites in relation to essential oil and curcumin accumulation in turmeric (*Curcuma longa* L.) *Photosynthetica* **40**: 109-113.

Dixon NE, Gazola C, Blakeley RL, Zerner B (1975). Jackbean urease (EC 3.5.1.5), a metalloenzyme. A simple biological role for nickel? *J. Am. Chem. Soc.* **97**: 4131-4133.

Domingo AL, Nagatomo Y, Tamai M, Takaki H (1992). Free tryptophan and indoleacetic acid in zinc-deficient radish shoots. *Soil Sci. Plant Nutr.* **38**: 261-267.

Donahue JL, Okpodu CM, Cramer CL, Grabau EA, Alscher RG (1997). Responses of antioxidants to paraquat in pea leaves. *Plant Phynol.* **113**: 249-257.

Donald CM, Prescott JA (1975). Trace elements in Australian crops and pasture production. In *Trace Elements in Soil Plant Animal System.* DJD Nicholas, AR Egan (eds) Academic Publishers, Sydney. 7-37.

Dong B, Rengel Z and Graham RD (1995). Characters of root geometry of wheat genotypes differing in zinc efficiency. *J. Plant Nutr.* **18**: 2761-2773.

Dordas C, Brown PH (2000). Permeability of boric acid across lipid bilayers and factors affecting it. *J. Membr. Biol.* **175**: 95-105.

Dordas C, Chrispeels MJ, Brown PH (2000). Permeability and channel-mediated transport of boric acid across membrane vesicles isolated from squash roots. *Plant Physiol.* **124**: 1349-1361.

Droppa M, Masojidik J, R'ozsa Z, Wolak A, Harvath LJ, Farkas T, Horvath G (1987). Characteristics of Cu deficiency induced inhibition of photosynthetic electron transport in spinach chloroplasts. *Biochem. Biophys. Acta.* **891**: 75-84.

Droppa M, Terry N, Horváth G (1984). Effect of copper deficiency on photosynthetic electron transport. *Proc. Natl. Acad. Sci.* (USA) **81**: 2369-2373.

Droux M, Miginiac-Maslow M, Jacquot J-P, Gadal P, Crawford NA (1987). Ferredoxin-thioredoxin reductase: A catalytically active dithiol group links photoreduced ferredoxin to thioredoxin fuctional in photosynthetic enzyme regulation. *Arch. Biochem. Biophys.* **256**: 372-380.

Durga D, Srinivasan PS, Balakreshnan K (1996). Influence of Zn, Fe and Mn on photosynthesis and yield of *Citrus sinensis*. Indian *J. Plant Physiol.* **2**: 174-176.

Dwivedi RS, Randhawa NS (1974). Evaluation of a rapid test for hidden hunger of zinc in plants. *Plant Soil* **40**: 445-451.

Dwivedi RS, Takkar PN (1974). Ribonuclease activity as an index of hidden hunger of zinc in crops. *Plant Soil* **40**: 173-181.

Eckhardt U, Margues AM, Buckhout TJ (2001). Two iron regulated cation transporters from tomato complement metal uptake deficient yeast mutants. *Plant Mol. Biol.* **45**: 437-448.

Edward GE, Mohamad AK (1973). Reduction in carbonic anhydrase activity in Zn deficient leaves of *Phaseolus vulgaris*. *Crop Sci.* **13**: 351-354.

Eide D, Broderius M, Feit J, Guerinot ML (1996). A novel iron-regulated metal transporter from plants identified by functional expression in yeast. *Proc. Natl. Acad. Sci.* USA. **93**: 5624-5628.

Eilers T, Schwarz G, Brinkmann H, Witt C, Richter T, Nieder J, Koch B, Hille R, Hansch R, Mendel RR (2001). Identification and biochemical characterization of *Arabidopsis thaliana* sulphite oxidase: a new player in plant sulphur metabolism. *J. Biol. Chem.* **276**: 46989-46994.

El-Shintinawy F (1999). Structural and functional damage caused by boron deficiency in sunflower leaves. *Photosynthetica.* **36**: 565-572.

Elstner EF (1991). Mechanism of oxygen activation in different compartments of plants In: *Active Oxygen Species, Oxidative Stress and Plant Metabolism*. Am. Soc. Plant Physiol. Rockville MD 13-25.

Ender Ch, Li MQ, Martin B, Povh B, Nobiling R, Reiss H-D, Traxel K (1983). Demonstration of polar zinc distribution in pollen tubes of *Lilium longiflorum* with the Hiedelberg proton microproble. *Protoplasma* **116**: 201-203.

Engelsma G (1972). The possible role of divalent magnaese ions in the photoinduction of phenylalanine ammonia lyase. *Plant Physiol.* **50**: 599-602.

Engvild KC (1986). Chlorine-containing natural compounds in higher plants. *Phytochemistry* **25**: 781-791.

Epstein E (1972). *Mineral Nutrition of Plants. Principles and Perspectives.* Wiley, New York.

Erenoglu B, Cakmak I, Marschner H, Romheld V, Eker S, Daghan H, Kalayci M, Ekiz H (1996). Phytosiderophare release does not relate well with zinc efficiency in different bread wheat genotypes. *J. Plant Nutr.* **19**: 1576-1580.

Espen DM, De Nisi P, Zocchi G (2000). Metabolic responses in cucumber (*Cucumis sativus* L.) roots under Fe-deficiency : A ^{31}P-NMR *in vivo* study. *Planta* **210**: 985-992.

Evans PT, Malmberg (1989). Do polyamines have roles in plant development. *Ann. Rev. Plant Mol. Biol.* **40**: 239-269.

Eyster C, Brown TE, Tanner HA, Hood SL (1958). Manganese requirement with respect to growth, Hill reaction and photosynthesis. *Plant Physiol.* **33**: 235-241.

Fackler U, Goldbach H, Weiler EW, Amberger A (1985). Influence of boron deficiency on indol 3-yl-acetic acid and abscisic acid levels in roots and shoot tips. *J. Plant Physiol.* **119**: 295-299.

Falchuk KH, Ulpino L, Mazus B, Valee BL (1977). *E. gracilis* RNA polymerase I: A zinc metalloenzyme. *Biochem. Biophys. Res. Commun.* **74**: 1206-1212.

Farwell AJ, Farina MPW, Channon P (1991). Soil acidity effects on premature germination in immature maize grain. In: *Plant-Soil interactions at low pH.* VC Baligor, RP Murrmann (eds) Kluwer Academic Publishers, Dordrecht. 355-361.

Federico R, Angelini R (1991). Polyamine catabolism in plants. In *Biochemistry and Physiology of Polyamines in Plants.* RD Slocum, HE Flores (eds) CRC Press. Boca Raton, FL: 41-56.

Felle HH (1994). The H^+/Cl^- symporter in root hair-cells of *Sinapis alba*. An electrophysiological study using ion-selective electrodes. *Plant Physiol.* **106**: 1131-1136.

Ferrol N, Belver A, Roldin M, Rodriguez-Rosales MP, Donaire JP (1993). Effects of boron on proton transport and membrane properties of sunflower (*Helianthus annuus* L.) cell microsomes. *Plant Physiol.* **103**: 763-769.

Ferrol N, Donaire JP (1992). Effect of boron on plasma membrane proton extrusion and redox activity in sunflower cells. *Plant Sci.* **86**: 41-47.

Feussner I, Wasternack C (2002). The lipoxygenase pathway. *Annu. Rev. Plant Biol.* **53**: 275-297.

Fido RJ, Gundry CS, Hewitt EJ, Notton BA (1977). Ultrastructural features of molybdenum deficiency and whiptail of cauliflower leaves. Effect of nitrogen source and tungsten substitution for molybdenum. *Aust. J. Plant Physiol.* **4**: 675-689.

Fischer ES, Thimm O, Rengel Z (1997). Zinc nutrition influences gas exchange in wheat. *Photosynthetica* **33**: 505-508.

Fleischer A, O'Neill MA, Ehwald R (1999). The pore size of nongraminaceous plant cell walls is rapidly decreased by borate ester cross-linking of the pectic polysacchaside rhamnogalacturonan II. *Plant Physiol.* **121**: 829-838.

Fleming BL (1995). Organochlorines in perspective. *Tappi J.* **78**: 93-98.

Flowers TJ (1988). Chloride as a nutrient and as an osmoticum. In: *Advances in Plant Nutrition.* B Tinker, A Läuchli (eds) Praeger, New York. 55-78.

Fobis-Loisy 1, Aussei L, Briat J-F (1996). Post-transcriptional regulation of plant ferritin accumulation in response to iron as observed in the maize mutant *ys1. FEBS Lett.* **397**: 149-154.

Fournier JM, Alcantara E, de la Guardia MD (1992). Organic acid accumulation in roots of two sunflower lines with a different response to iron deficiency. J. Plant Nutr. **15**: 1747-1755.

Fox TC, Guerinot ML (1998). Molecular biology of cation transport in plants. *Annu. Rev. Plant Physiol. Plant Mol. Biol.* **49**: 669-696.

Foy CD (1983). Plant adaptation to mineral stress in problem soils. *Iowa State J. Res.* **57**: 355-391.

Franco AA, Munns DN (1981). Response of *Phaseolus vulgaris* L. to molybdenum under acid conditions. *Soil Sci. Soc. Am. J.* **45**:1144-1148.

Franklin-Tong VE (1999). Signalling and the modulation of pollen tube growth. *Plant Cell* **11**: 727-738.

Fridovich I (1986). Biological effects of the superoxide radical. *Arch. Biochem. Biophys.* **247**: 1-9.

Fridovich I (1986). Superoxide dismutases. *Adv. Enzymology* **41**: 35.

Fromm J, Eschrich W (1988). Transport processes in stimulated and non-stimulated leaves of *Mimosa pudica* III. Displacement of ions during seismonastic leaf movements. *Trees* **2**: 65-72.

Fromm J, Eschrich W (1989). Correlations of ionic movements with phloem loading and unloading in barley leaves. *Plant Physiol. Biochem.* **27**: 577-588.

Frono DA, Yoshida S, Asher CJ (1975). Zinc deficiency in rice I Soil factors associated with the deficiency. *Plant Soil* **42**: 537-550.

Fujiwara A, Tsutsumi M (1955). Biochemical studies of microelements in green plants. I Deficiency symptoms of microelements on barley plants and changes in indoleacetic acid oxidase (in Japanese). *J. Sci. Soil Tokyo* **26**: 259-262.

Fujiwara A, Tsutsumi M (1962). Biochemical studies of micronutrients in green plants IV. Status of chloroplasts and rate of photosynthesis in micronutrient deficient barley leaves. *Tohoku J. Agric. Res.* Japan **13**: 169-174.

Gaiser JC, Robinson-Beers K, Gasser CS (1995). The *Arabidopsis SUPER-MAN* gene mediates asymmetric growth of the outer integument of ovules. *Plant Cell.* **7**: 333-345.

Galloway CM, Dugger WM (1990). Boron inhibition of phosphoglucomutase from peas. *J. Plant Nutr.* **13**: 817-825.

Garcia-González M, Mateo P, Bonilla I (1988). Boron protection for O_2 diffusion in heterocyst of *Anabaena* sp. PCC 7119. *Plant Physiol.* **87**: 785-789.

Garcia-González M, Mateo P, Bonilla I (1991). Boron requirement for envelope structure and function in *Anabaena* PCC 7119 heterocysts. *J. Exp. Bot.* **42**: 425-429.

Garde J, Kinghorn JR, Tomsett AB (1995). Site directed mutagenesis of nitrate reductase from *Aspergillus nidulans. J. Biol. Chem.* **270**: 6644-6650.

Garg OK, Sharma AN, Kona GRSS (1979). Effect of boron on the pollen vitality and yield of rice plant (*Oryza sativa* L. var. Jaya). *Plant Soil* **52**: 591-594.

Gauch HG, Dugger WM Jr (1954). The physiological action of boron in higher plants: A review and interpretation: Univ. Maryland. Agric. Exp. Stn. Bull. A-80 (Technical), College Park, Maryland.

Gene Y, Mc Donald GK, Graham RD (2002). Critical deficiency concentration of zinc in barley genotypes differing in zinc efficiency and its relation to growth responses. *J. Plant Nutr.* **25**: 545-560.

Georgatsou E, Mavrogiannis LA, Fragiadakis GS, Alexandraki D (1997). The yeast Fre1p /Fre2p cupric reductases facilitate copper uptake and are regulated by the copper modulated Mac1p activator. *J. Biol. Chem.* **272**: 13786-13792.

George GN, Mertens JA, Campbell WH (1999). Structural changes induced by catalytic-turnover of the molybdenum site of *Arabidopsis* nitrate reductase. *J. Amer. Chem. Soc.* **121**: 9730-9731.

Gerendas J, Polacco JC, Freyermuth SK, Sattelmacher B (1999). Significance of nickel for plant growth and metabolism. *J. Plant Nutr. Soil Sci.* **162**: 241-256.

Gerendas J, Sattelmacher B (1997). Significance of Ni supply for growth, urease activity and the contents of urea, amino acids, and mineral nutrients of urea-grown plants. *Plant Soil* **190**: 153-162.

Ghildiyal MC, Pandey M, Sirohi GS (1977). Proline accumulation under zinc deficiency in mustard. *Curr. Sci.* **46**: 792-794.

Ghildiyal MC, Tomar OPS, Sirohi GS (1978). Response of chickpea genotypes to zinc in relation to photosynthesis, nodulation and dry-matter distribution. *Plant Soil.* **49**: 505-516.

Ghoneim MF, Bussler W (1980). Diagnosis of zinc deficiency in cotton. Z. *Pflanzenernahr Bodenkd.* **143**: 377-384.

Gibbs J, Greenway H (2003). Mechanism of anoxia tolerance in plants I growth, survival and anaerobic catabolism. *Fun.Plant Biol.* **30**:1-47.

Gibson TS, Leece DR (1981). Estimation of physiologically active zinc in maize by biochemical assay. *Plant Soil* **63**: 395-406.

Gold MH, Alic M (1993). Molecular biology of the lignin-degrading basidiomycete *Phanerochaete chrysoporium*. *Microbiol. Rev.* **57**: 605-622.

Goldbach H (1985). Influence of boron nutrition on net uptake and efflux of ^{32}P and ^{14}C glucose in *Helianthus annuus* roots and cell cultures of *Daucus carota*. *J. Plant Physiol.* **118**: 431-438.

Goldbach H (1997). A critical review of current hypotheses concerning the role of boron in higher plants: Suggestions for further research and methodological requirements. *J. Trace Microprobe Tech.* **15**: 51-91.

Goldberg S, Forster HS (1991). Boron sorption on calcareous soils and reference calcites. *Soil Sci.* **152**: 304-310.

Goldburg S (1997). Reactions of boron with soils. *Plant Soil.* **193**: 35-48.

Gomez-Rodriguez MV, Castillo JLD, Alvarez-Tinaut MC (1987). The evolution of glucose-6 P-dehydrogenase and 6 P-gluconate dehydrogenase activities and the ortho-diphenolic content of sunflower leaves cultivated under different boron treatments. *J. Plant Nutr.* **10**: 2211-2229.

Goodall DW and Gregory FG (1947). *Chemical Composition of Plants as an Index of their Nutritional Status*. Tech. Commun. No. 17. Imp. Bur. Hort. Plantn. Crops. East Malling, England.

Gordon WR, Schwemmer SS, Hillman WS (1978). Nickel and the metabolism of urea by *Lemna paucicostata* Hegelm 6746. *Planta* **140**: 265-268.

Gorham J (1995). Betaines in higher plants-biosynthesis and role in stress metabolism. In: *Aminoacids and their Derivatives in Higher Plants* (RM Wallsgrove, ed.) Cambridge University Press. Cambridge: 171-203.

Graham MJ, Stephens PA, Widholm JM, Nickell CD (1992). Soybean genotype evaluation for iron deficiency chlorosis using sodium bicarbonate and tissue culture. *J. Plant Nutr.* **15**: 1215-1225.

Graham RD (1975). Male sterility in wheat plants deficient in copper. *Nature* (London) **254**: 514-515.

Graham RD (1976). Anomalous water relations in copper deficient wheat plants. *Aust. J. Plant Physiol.* **3**: 229-236.

Graham RD (1980). The distribution of copper and soluble carbohydrates in wheat plants grown at high and low levels of copper supply. *Z. Pflanzenernahr. Bodenk.* **143**: 161-164.

Graham RD (1983). Effect of nutrient stress on susceptibility of plants to disease with particular reference to trace elements. *Adv. Bot. Res.* **10**: 221-276.

Graham RD (1988). Genotypic differences in tolerance to manganese deficiency. In *Manganese in Soils and Plants*. RD Graham, RJ Hannam, NC Uren (eds) Kluwer Academic, Dordrecht. 261-276.

Graham RD, Ascher JS, Ellis PAE, Shephard KW (1987a). Transfer to wheat of the copper efficiency factor carried on rye chromosome arm 5RL. *Plant Soil.* **99**: 107-114.

Graham RD, Ascher JS, Hynes SC (1992). Selecting zinc-efficient cereal genotypes for soils of low zinc status. *Plant Soil* **146**: 241-250.

Graham RD, Hannam RJ, Uren NC (eds) (1988). *Manganese in Soils and Plants*. Kluwer Academic Publishers, Dordrecht.

Graham RD, Pearce DT (1979). The sensitivity of hexaploid and octaploid triticales and their parent species to copper deficiency. *Aust. J. Agric. Res.* **30**: 791-799.

Graham RD, Rengel Z (1993). Genotypic variation in zinc uptake and utilization by plants. In *Zinc in Soils and Plants*. A.D. Robson (ed.), Kluwer Academic Publishers, Dordrecht. 107-118

Graham RD, Rovira AD (1984). A role of manganese in the resistance of wheat plants to take-all. *Plant Soil* **78**: 441-444.

Graham RD, Webb MJ (1991). Micronutrients and plant disease resistance and tolerance in plants. In *Micronutrients in Agriculture*. JJ Mortvedt, FR Cox, LM Shuman, RM Welch (eds) Soil Sci. Soc. Amer. Book Ser. No. 4. Madison. WI. 329-370.

Graham RD, Welch RM (1996). Breeding for staple food crops with high micronutrient density. Working Papers on Agricultural strategies for Micronutrients No. 3. *International Food Policy Research Institute*, Washington DC.

Graves CJ, Sutcliffe JF (1974). An effect of copper deficiency on the initiation and development of flower buds of *Chrysanthemum morifolium* grown in solution culture. *Ann. Bot.* **38**: 729-738.

Graziano M, Beligni MV, Lamattina L (2002). Nitric oxide improves internal iron availability in plants. *Plant Physiol.* **130**: 1852-1859.

Greenway H (1965). Plant responses to saline substrates. IV. Chloride uptake by *Hordeum vulgare* as affected by inhibitors, transpiration, and nutrients in the medium. *Aust. J. Biol. Sci.* **18**: 249-268.

Grewal HS , Graham RD (1997). Seed-Zn content influences early vegetative growth and zinc uptake in oil seed rape (*Brassica napus* and *Brassica juncea*) genotypes on zinc-deficient soils. *Plant Soil*.**129**:191-197.

Grewal HS, Graham RD, Rengel Z (1996). Genotypic variation in zinc efficiency and resistance to crown rot disease (*Fusarium graminearum* Schn. Group I) in wheat. *Plant Soil* **186**: 219-226.

Grewal HS, Stangoulis JCR, Potter TD, Graham RD (1997). Zinc efficiency of oilseed rape (*Brassica napus* and *B. juncea*) genotypes. *Plant Soil* **191**: 123-132.

Grigg JL (1953). Determination of the available molybdenum of soils. *New Zealand J. Sci. Technol.* **A34**: 405-414.

Grigg JL (1960). The distribution of molybdenum in soil of New Zealand. I Soils of the Northern Island. *New Zealand J. Agric. Res.* **3**: 69-86.

Grossniklaus U, Vielle-Calzada J-P, Hoeppner, MA, Gaglianio WB (1998). Maternal control of embryogenesis by *MEDEA*, a Polycomb-group gene in *Arabidopsis.Science*. **280**: 446-450.

Grotz N, Fox T, Connolly E, Park W, Guerinot ML, Eide D (1998). Identification of a family of zinc transporter genes from *Arabidopsis* that respond to zinc deficiency. *Proc. Natl. Acad. Sci.* USA **95**: 7220-7224.

Gruber B, Kosegarten H (2002). Depressed growth of non-chlorotic vine grown in calcareous soil is an iron deficiency symptom prior to leaf chlorosis. *J. Plant Nutr. Soil Sci.* 165 : 111-117.

Gruhn K (1961). Einfluss einer Molybdän-Düngung auf einige Stickstoff Fraktionen von Luzerne und Rotklee. *Z. Pflanzenernaehr, Dueng, Bodenkd.* **95**: 110-118.

Grusak MA (1995). Whole-root iron (III)-reductase activity throughout the life cycle of iron-grown *Pisum sativum* L. (Fabiaceae) : Relevence to the nutrition of developing seed. *Planta* **197**: 111-117.

Grusak MA, Pezeshgi S (1996). Shoot-to-Root signal transmission regulates root Fe (III) reductase activity in the *dgl* mutant of pea. *Plant Physiol.* **110**: 329-334.

Guan X, Stege J, Kim M, Dahmani Z, Fan N, Heifetz P, Barbas III CF, Briggs SP (2002). Heritable endogenous gene regulation in plants with designed polydactyl zinc finger transcription factors. *Proc. Natl. Acad. Sci.* USA. **99**: 13296-13301.

Gubler WD, Grogan RG, Osterli PP (1982). Yellows of mellons caused by molybdenum deficiency in acid soil. *Plant Dis.* **66**: 449-451.

Gueren J, Felle H, Matthieu Y, Kurkdjian A (1991). Regulation of intracellular pH in plant cells. *Int. Rev. Cytol.* **127**: 111-1173.

Guerinot ML (2000). The ZIP family of metal transporters. *Biochem. Biophys. Acta.* **1465**: 190-198.

Gupta AS, Webb RP, Holaday AS, Allen RD (1993). Overexpression of superoxide dismutase protects plants from oxidative stress: induction of ascorbate peroxidase in superoxide dismutase-overexpressing plants. *Plant Physiol.* **103**: 1067-1073.

Gupta UC (ed.) (1997a). *Molybdenum in Agriculture*. Cambridge University Press. New York.

Gupta UC (1997b). Soil and plant factors affecting molybdenum uptake by plants. In: *Molybdenum in Agriculture.* UC Gupta (ed.) Cambridge University Press, New York. 71-91.

Gupta UC (1997c). Deficient, sufficient and toxic concentration of molybdenum in crops. In: *Molybdenum in Agriculture.* UC Gupta (ed.) Cambridge University Press, New York. 150-159.

Gupta UC (1997d). Symptoms of molybdenum deficiency and toxicity in crops. In: *Molybdenum in Agriculture.* UC Gupta (ed.) Cambridge University Press, New York. 160-170.

Gupta UC (ed.) (1993). *Boron and its Role in Crop Production*. CRC, Press Boca Raton Fl.

Gupta UC, Lipsett J (1981). Molybdenum in soils, plants and animals. *Adv. Agron.* **34**: 73-115.

Gupta UC, McKay DC (1965). Extraction of water soluble copper and molybdenum from podzol soils. *Proc. Soil Sci. Soc. Amer.* **29**: 323.

Gurley WB, Giddens J (1969). Factors affecting uptake, yield response and carry over of molybdenum on soybean seed. *Agron J.* **61**: 7-9.

Guss JM, Merritt EA, Phizackerley RP, Freeman HC (1996). The structure of a phytocyanin, the basic blue protein from cucumber, redefined at 1.8 Å resolution. *J. Mol. Biol.* **262**: 686-705.

Hacisalihoglu G, Hart JJ, Wang Y-H, Cakmak I, Kochian LV (2003). Zinc efficiency is correlated with enhanced expression and activity of zinc-requiring enzymes in wheat. *Plant Physiol.* **131**: 595-602.

Hafner H, Ndunguru BJ, Bationo A, Marschner H (1992). Effect of nitrogen, phosphorus and molybdenum application on growth and symbiotic N_2-fixation of groundnut in an acid sandy soil in Niger. *Fert. Res.* **31**: 69-77.

Hager A, Helmle M (1981). Properties of an ATP-fueled Cl⁻-dependent proton pump localized in membranes of microsomal vesicles from maize coleoptiles. *Z. Naturforsch., Biosci.* C: **36**: 997-1008.

Hakisalihoglu G, Hart JJ, Kochian LV (2001). High and Low affinity zinc transporter systems and their possible role in zinc efficiency in bread wheat. *Plant Physiol.* **125**: 456-463.

Hall JL, Williams LE (2003). Transition metal transporters in plants. *J. Exp. Bot.* **54**: 2601-2613.

Halliwell B, Gutteridge JMC (1986). Iron and free radical reactions: Two aspects of antioxidant protection. *Trends Biochem. Sci.* **11**: 372-375.

Halliwell B, Gutteridge JMC (1989). *Free Radicals in Biology and Medicine* 2ⁿᵈ ed. Clarendon Press, Oxford.

Hansen NC, Jolley VD (1995). Phytosiderophore release as a criterion for genotypic evaluation of iron efficiency in oat. *J. Plant Nutr.* **18**: 455-465.

Hansen NC, Jolley VD, Berg WA, Hodges ME, Krenzer EG (1996). Phytosiderophore release related to susceptibility of wheat to iron deficiency. *Crop Science* **36**: 1473-1476.

Hanson E, Breen P (1985). Effects of fall boron sprays and environmental factors on fruit set and boron accumulation in 'Italian' prune flowers. *J. Amer. Soc. Horti. Sci.* **110**: 389-392.

Hanson EJ (1991). Movements of boron out of tree fruit leaves. *HortScience* **26**: 271-273.

Hanstein SM, Felle HH (2002). CO_2-triggered chloride release from guard cells in intact fava bean leaves. Kinetics of the onset of stomatal closure. *Plant Physiol.* **130**: 940-950.

Harms K, Atzorn R, Brash A, Kühn H, Wasternnack C, Willmitzer L, Pena-Cortes H (1995). Expression of a flax allene oxide synthase cDNA leads to increased endogenous jasmonic acid (JA) levels in transgeinc potato plants but not to corresponding activation of JA-responding genes. *Plant Cell* **7**: 1645-1654.

Harrison R (2002). Structure and function of xanthine oxidoreductase: Where are we now ? *Free Rad. Biol. Med.* **33**: 774-797.

Hart JJ, Norvell WA, Welch RM, Sullivan LA, Kochian LV (1998). Characterization of zinc uptake, binding, and translocation in intact seedlings of bread and durum wheat cultivars. *Plant Physiol.* **118**: 219-226.

Hartfield R, Vermerris W (2001). Lignin formation in plants. The dillemma of linkage specificity. *Plant Physiol.* **126**: 1351-1357.

Hasett R, Kosman DJ (1995). Evidence for Cu (II) reduction as a component of copper uptake by *Saccharomyces cerevisiae. J. Biol. Chem.* **270**: 128-134.

Hashimoto K, Yamasaki S (1976). Effects of molybdenum application on the yield, nitrogen nutrition and nodule development of soybeans. *Soil Sci. Plant Nutr.* **22**: 435-443.

Haslett BS, Reid RJ, Rengel Z (2001). Zinc mobility in wheat : Uptake and distribution of zinc applied to leaves or roots. *Ann. Bot.* **87**: 379-386.

Hatch MD, Burnell JN (1990). Carbonic anhydrase activity in leaves and its role in the first step of C_4 photosynthesis. *Plant Physiol.* **93**: 825-828.

Havlin JL, Soltanpour PN (1981). Evaluation of the NH_4HCO_3-DTPA soil test for iron and zinc. *Soil Sci. Soc. Amer. J.* **45**: 70-75.

Hedden P, Phillips AL (2000). Gibberellin metabolism: new insights revealed by the genes. *Trends Plant Sci.* **5**: 523-530.

Heenan DP, Campbell LC (1990). The influence of temperature on the accumulation and distribution of manganese in two cultivars of soybean (*Glycine max*(L) Mur). *Aust. J. Agric. Res.* **41**: 835-843.

Hell R, Stephan UW (2003). Iron uptake, trafficking and the homeostasis in plants. *Planta.* **216**: 541-551.

Helliwell CA, Chandler PM, Poole A, Dennis ES, Peacock WJ (2001). The CYP88A cytochrome P450, *ent*-kaurenoic acid oxidase, catalyzes three steps of the gibberellin biosynthesis pathway. *Proc. Natl. Acad. Sci. USA* **98**: 2065-2070.

Henriques FS (1989). Effects of copper deficiency on the photosynthetic apparatus of sugarbeet (*Beta vulgaris* L.). *J. Plant Physiol.* **135**: 453-458.

Henriques FS (2001). Loss of blade photosynthetic area and of chloroplasts photochemical capacity account for reduced CO_2 assimilation rates in zinc-deficient sugarbeet leaves. *J. Plant Physiol.* **158**: 915-919.

Henriques R, Jasik J, Klein M, Martinoia E, Feller U, Schell J, Pais MS, Koncz C. (2002). Knock out of *Arabidopsis* metal transporter gene *IRT1* results in iron deficiency accompanied by cell differentiation defects. *Plant Mol. Biol.* **50**: 587-597.

Herbik A, Koch G, Mock H-P, Dushkov D, Czihal A, Thielmann J, Stephan UW, HB (1999). Isolation, characterization and c DNA cloning of nicotianamine synthase from barley: A key enzyme for iron homeostasis in plants. *Eur. J. Biochem.* **265**: 231-239.

Herman PL, Ramberg H, Baack RD, Markwell J, Osterman JC (2002). Formate dehydrogenase in *Arabidiopsis thaliana*. Overexpression and subcellular localization in beans. *Plant Sci.* **163**: 1137-1145.

208

Herrmann KM (1995). The shikimate pathway: Early steps in the biosynthesis of aromatic compounds. *Plant Cell* **7**: 909-919.

Herrmann KM, Weaver LM (1999). The shikimate pathway. *Annu. Rev, Plant Physiol. Plant Mol. Biol.* **50**: 473-503.

Heslop-Harrison JS, Roger BJ (1986). Chloride and potassium ions and turgidity in the grass stigma. *J. Plant Physiol.* **124**: 55-60.

Hether NH, Olsen RA, Jackson LL (1984). Clinical identification of iron reductants exuded by plant roots. *J. Plant Nutr.* **7**: 667-676.

Heuwinkel H, Kirkby EA, Le Bot J, Marschner H (1992). Phosphorus deficiency enhances molybdenum uptake by tomato plants. *J. Plant Nutr.* **15**: 549-568.

Hewitt EJ (1956). Symptoms of molybdenum deficiency in plants. *Soil Sci.* **81**: 159-171.

Hewitt EJ (1963). The essential nutrient elements : Requirements and interactions in plants. In *Plant Physiology* Vol. 3. FC Steward (ed.) Academic Press, New York. 137-360.

Hewitt EJ (1966). *Sand and Water Culture Methods Used in the Study of Plant Nutrition* (2nd ed.) Tech. Commun. No. 22. Commonw. Agric. Bureaux Farnham Royal. Bucks, England.

Hewitt EJ, Agarwala SC, Jones EW (1950). Effect of molybdenum status on the ascorbic acid content of plants in sand culture. *Nature* **166**: 1119-1121.

Hewitt EJ, Agarwala SC, Williams (1957). Molybdenum as a plant nutrient. VIII. The effect of different molybdenum levels and nitrogen supplies on the nitrogen fractions in cauliflower plants grown in sand culture. *J. Hort. Sci.* **32**: 34-48.

Hewitt EJ, Bond G (1966). The cobalt requirement of non-legume root nodule plants. *J. Exp. Bot.* **17**: 480-491.

Hewitt EJ, Gundry CS (1970). The molybdenum requirement of plants in relation to nitrogen supply. *J. Hort. Sci.* **45**: 351-358.

Hewitt EJ, McCredy CC (1956). Molybdenum as a plant nutrient VII. The effects of different molybdenum and nitrogen supplies on yields and composition of tomato plants grown in sand culture. *J. Hortic. Sci.* **31**: 284-290.

Hewitt EJ, Tatham P (1960). Interaction of mineral deficiency and nitrogen supply on acid phosphatase activity in leaf extracts. *J. Exp. Bot.* **11**: 367-376.

Higuchi K, Kanazawa K, Nishizawa NK, Chino M, Mori S (1994). Purification and characterization of nicotianamine synthase from Fe-deficient barley roots. *Plant Soil* **165**: 173-179.

Higuchi K, Suzuki K, Nakanishi H, Yamaguchi H, Nishizawa NK, Mori S (1999). Cloning of nicotinamine synthase genes, novel genes involved in the biosynthesis of phytosiderophores. *Plant Physiol.* **119**: 471-480.

Hill J, Robson AD, Loneragan, JF (1979). The effects of copper supply and shading on copper retranslocation from old leaves. *Ann. Bot.* **43**: 449-457.

Hille R (1996). The mononuclear molybdenum enzymes. *Chem. Rev.* **96**: 2757-2816.

Himelblau E, Mira M, Lin S-J, Culotta VC, Penarrubia L, Amasino RM. (1998). Identification of a functional homolog of the yeast copper homeostasis gene *ATX1* from *Arabidopsis. Plant Physiol.* **117**: 1227-1234.

Hinde RW and Finch LR (1966). The activities of phosphatase, pyrophosphatases and adenosine triphosphatases from normal and boron deficient bean roots. *Phytochemistry* **5**: 619-624.

Hirsch AM, Pengelly WL, Torrey JG (1982). Endogenous IAA levels in boron-deficient and control root tips of sunflower *Bot. Gaz.* (Chicago): **143**: 15-19.

Hirsch AM, Torrey JG (1980). Ultrastructural changes in sunflower root cells in relation to boron deficiency and added auxin. *Can J. Bot.* **58**: 856-866.

Hirschi KD, Korenkov VD, Wilganowski NL, Wagner GJ (2000). Expression of *Arabidopsis CAX2* in tobacco. Altered metal accumulation and increased manganese tolerance. *Plant Physiol.* **124**: 125-134.

Hocking PJ (1980). The composition of phloem exudate and xylem sap from tree tobacco (*Nicotiana glauca* Groh.) *Ann. Bot.* **45**: 633-643.

Höfner W, Grieb R (1979). Einfluss von Fe-und Mo-Mangel auf den Ionengehalt mono-und dikotyler Pflanzen unterschiedlicher Chloroseanfälligkeit. *Z. Pflanzenernähr. Badenk.* **142**: 626-638.

Hoganson CW, Babcock GT (1997). A metalloradical mechanism for the generation of oxygen from water in photosynthesis. *Science* **277**: 1953-1956.

Hoganson CW, Casey PA, Hansson Ö (1993). Flash photolysis studies of manganese-depleted photosystem II : Evidence for binding of Mn^{2+} and other transition metal-ions. *Biochem. Biophys. Acta.* **1057**: 399-406.

Hopkins BD, Whitney DA, Lamond RE, Jolley VD (1998). Phytosiderophore release by sorghum, wheat and corn under Zn deficiency. *J. Plant Nutr.* **21**: 2623-2637.

Hopkins MF (1983). Metals and photosynthesis. In *Metals and Micronutrients: Uptake and Utilization by Plants*. DA Robb, WS Pierpoint (eds) Academic Press, New York: 147-168.

Hopmans P (1990). Stem deformity in *Pinus radiata* plantations in South-Eastern Australia I. Response to copper fertilization. *Plant Soil* **122**: 97-104.

Horváth G, Droppa M, Terry N (1983). Changes in photosynthetic attributes in response to copper depletion in sugarbeets (*Beta vulgaris* L.). *J. Plant Nutr.* **6**: 971-981.

Horváth G, Melis A, Hideg E, Droppa M, Vigh I (1987). Role of lipids in the organization of photosystem II studied by homogenous catalytic hydrogenation of thylakoid membranes in situ. *Biochem. Biophys. Acta* **891**: 68-71.

Hossain B, Hiata N, Nagatomo Y, Suiko M, Takaki H (1998). Zinc nutrition and levels of endogenous Indole-3-acetic acid (IAA) in radish shoots. *J. Plant Nutr.* **21**: 1113-1128.

Houston-Cabassa C, Ambard-Bretteville F, Moreau F, de Virville JD, Rémy R, Cola des Frances-c Small C (1998). Stress induction of mitochondrial formate dehydrogenase in potato leaves. *Plant Physiol.* **116**: 626-635.

Howe GA, Lee GI, Itoh A, Li L, De-Rocher AE (2000). Cytochrome P450 dependent metabolism of oxylipins in tomato. Cloning and expression of allene oxide synthase and fatty acid hydroperoxide lyase. *Plant Physol.* **123**: 711-724.

Hsu W, Miller GW (1968). Iron in relation to aconitate hydratase activity in *Glycine max* Merr. *Biochim. Biophys. Acta.* **151**: 711-713.

Hu H, Brown PH (1994). Localization of boron in cell walls of squash and tobacco and its association with pectin-evidence for a structural role of boron in the cell wall. *Plant Physiol.* **105**: 681-689.

Hu H, Brown PH (1996) Phloem mobility of boron is species dependent: Evidence for boron mobility in sorbitol-rich species. *Ann. Bot.* **77**: 497-505.

Hu H, Brown PH (1997). Absorption of boron by plant roots. *Plant Soil.* **193**: 49-58.

Hu H, Brown PH, Labavitch JM (1996). Species variability in boron requirement is correlated with cell wall pectin. *J. Exp. Bot.* **47**: 227-232.

Hu H, Penn SG, Lebrilla CB, Brown PH (1997) Isolation and characterization of soluble boron complexes in higher plants. *Plant Physiol.* **113**: 649-655.

Hu H, Sparks D (1991). Zinc deficiency inhibits chlorophyll synthesis and gas exchange in 'Stuart' pecan. *Hort. Sci.* **26**: 267-268.

Huang C, Webb AJ, Graham RD (1996). Pot size affects expression of Mn efficiency in barley. *Plant Soil* **178**: 205-208.

Huang C, Webb MJ, Graham RD (1993). Effect of pH on Mn absorption by barley genotypes in a chelate-buffered nutrient solution. *Plant Soil* **155/158**: 437-440.

Huang I-J, Welkic GW, Miller GW (1992). Ferredoxin and flavodoxin analysis in tobacco in response to iron stress. *J. Plant Nutr.* **15**: 1765-1782.

Huang L, Pant J, Bell RW, Dell B, Deane K (1996). Effect of boron deficiency and low temperatures on wheat sterility In *Sterility in Wheat in Subtropical Asia: Extent, Causes and Solutions*. HM Rawson, KD Subedi (eds. *Proceeding No 72 ACIAR* Canbera No 72:91-101.

Huang L, Pant J, Dell B, Bell RW (2000). Effect of boron deficiency on anther development and floral fertility in wheat (*Triticum aestivum* L. 'Wilgoyne') *Ann. Bot.* **85**: 493-500.

Huang, C, Webb MJ, Graham RD (1994). Manganese efficiency is expressed in barley growing in soil system but not in a solution culture. *J. Plant Nutr.* **17**: 83-95.

Huber DM, Graham RD (1999). The role of nutrition in crop resistance and tolerance to diseases. In *Mineral Nutrition of Crops. Fundamental Mechanisms and Implications.* Z Rengel (ed.) Haworth Press. New York: 169-204.

Huber DM, McCay-Buis TS (1993). A multiple component analysis of the take-all disease of cereals. *Plant Disease.* **77**: 437-449.

Huber DM, Wilhelm NS (1988). The role of manganese in resistance to plant diseases. In *Manganese in Soils and Plants.* RD Graham, PJ Hannam, NC Uren (eds) Kluwer Academic Publishers, Dordrecht. 155-173

Huffman DL, O' Halloran TV (2001). Function, structure and mechanism of intracellular trafficking proteins. *Annu. Rev. Biochem.* **70**: 677-701.

Hughes NP, Williams RJP (1988). An introduction to manganese biological chemistry. In *Manganese in Soils and Plants.* RD Graham, RJ Hannam, NC Uren (eds). Kluwer Academic, Dordrecht. 7-19.

Hull AK, Vij R, Celenza JL (2000). *Arabidopsis* cytochrome P 450s that catalyze the first step of tryptophan–dependent indole-3-acetic acid synthesis. *Proc. Natl. Acad. Sci.* USA. **97**:2379-2384.

Ishii T, Matsunaga T (1996) Isolation and characterization of a boron-rhamnogalacturonan-II complex from cell walls of sugarbeet pulp. *Carbohydr. Res.* **284**: 1-9.

Ishii T, Matsunaga T, Iwai H, Satoh H, Taushita J (2002). Germanium does not substitute for boron in cross-linking of rhamnogalacturonan II in pumpkin cell walls. *Plant Physiol.* **130**: 1967-1973.

Ishizuka J (1982). Characterization of molybdenum absorption and translocation in soybean plants *Soil Sci. Plant Nutr.* **28**: 63-78.

Iturbe-Ormaetxe I, Moran JF, Arrese-Igor C, Gogorcena Y, Klucas RV, Becana M (1995) Activated oxygen and antioxidant defenses in iron-deficient pea plants. *Plant Cell Environ.* **18**: 421-429.

Jarvis SC (1981). Copper concentration in plant and their relationship to soil properties. In *Copper in Soils and Plants.* JF Loneragan, AD Robson, RD Graham (eds) Academic Press, London. 265-285

Jeffrey AJ, McCallum LE (1988) Investigation of a hot 0.01$\underline{M}$ calcium chloride soil boron extraction procedure followed by ICP-AES analysis. *Commun. Soil Sci. Plant Anal.* **19**: 663-673.

Jeshke WD, Pate JS (1991). Cation and chloride partitioning through xylem and phloem within the whole plant *Ricinus communis* L. under conditions of salt stress. *J. Plant Physiol.* **132**: 45-53.

Jeshke WD, Wolf O, Hartung W (1992). Effect of NaCl salinity on flows and partitioning of C, N and mineral ions in whole plants of white lupin, *Lupinus albus* L. *J. Exp. Bot.* **43**: 777-788.

Jewell AW, Murry BG, Alloway BJ (1988). Light and electron microscopic studies on pollen developments in barley (*Hordeum vulgare* L.) grown under copper-sufficient and deficient conditions. *Plant Cell Environ.* **11**: 237-281.

Johnson D, Dear DR, Smith AD, Johnson MK (2005). Structure, function and formation of biological iron-sulfur clusters. *Annu. Rev. Biochem.* 74: 247-281.

Johnson DL, Albert LS (1967) Effect of selected nitrogen bases and boron on the ribonucleic acid content, elongation and visible deficiency symptoms in tomato root tips. *Plant Physiol.* 42: 1307-1309.

Johnson AD, Simons JG (1979). Diagnostic indices of zinc deficiency in tropical legumes. *J. Plant Nutr.* 1: 123-149.

Johnson C, Kerridge PC, Sultana A (1997). Response of forage legumes and grasses to molybdenum. In *Molybdenum in Agriculture*. UC Gupta (ed.). Cambridge University Press, New York: 202-228.

Johnson CM, Stout PR, Broyer TC, Carlton AB (1957). Comparative chlorine requirements of different plant species. *Plant Soil* 8: 337-353.

Johnson GV, Fixen PE (1990). Testing soils for sulphur, boron, molybdenum and chlorine. In: *Soil Testing and Plant Analysis* 3rd ed. RL Westerman (ed.) *Soil Sci. Soc. Amer.* Madison, WI. 265-273.

Johnson GV, Mayeux PA, Evans HJ (1966). A cobalt requirement for symbiotic growth of *Azolla filiculoides* in the absence of combined nitrogen. *Plant Physiol.* 41: 852-855.

Jolivet P, Bergeron E, Meunier J-C (1995). Evidence for sulphite oxidase activity in spinach leaves. *Phytochemistry* 40: 667-672.

Jolley VD, Brown JC (1987). Soybean response to iron-deficiency stress as related to iron supply in the growth medium. *J. Plant Nutr.* 10: 637-651.

Jolley VD, Brown JC (1991a). Differential response of Fe-efficient corn and Fe-inefficient corn and oat to phytosiderophore released by Fe-efficient *Coker 227* oat. *J. Plant Nutr.* 14: 45-58.

Jolley VD, Brown JC (1991b). Factors in iron stress response mechanism enhanced by Zn deficiency stress in Sanilac, but not Saginaw navy bean. *J. Plant Nutr.* 14: 257-265.

Jolley VD, Brown JC, Nugent PE (1991). A genetic related response to iron deficiency stress in muskmelon. *Plant Soil.* 130: 87-92.

Jolley VD, Cook KA, Hausen NC, Stevens WB (1996). Plant Physiological responses for genotypic evaluation of iron efficiency in strategy I and strategy II plants. A review. *J. Plant Nutr.* 19: 1241-1255.

Jolley VD, Fairbanks DJ, Stevens WB, Terry RE, Orf JH (1992). Root iron reduction capacity for genotypic evaluation of iron efficiency in soybean. *J. Plant Nutr.* 15: 1679-1690.

Jones JB jr (1991). Plant tissue analysis in micronutrients. In *Micronutrients in Agriculture* 2nd Ed. JJ Mortvedt, FR Cox, LM Schuman, RM Welch (eds) Soil Sci. Soc. Amer. Book Ser. 4. Madison WI. 477-521.

Jones JB jr (1992). Sample preparation and determination of iron in plant tissue samples. *J. Plant Nutr.* 15: 2085-2108.

Jones JB Jr. (1988). Soil test methods: Past present and future use of soil extractants. *Commun. Soil Sci. Plant Anal.* 29: 1543-1552.

Jongruayzup S, Dell B, Bell RB (1994). Distribution and redistribution of molybdenum in black gram (*Vigna mungo* L. Hepper) in relation to molybdenum supply. *Ann. Bot.* **73**: 161-167.

Jordan A, Reichard P (1998). Ribonucleotide reductases. *Annu. Rev. Biochem.* **67**: 71-98.

Judel GK (1972). Effect of copper and nitrogen deficiency on phenol oxidase activity and content of phenols in leaves of sunflower (*Helianthus annuus*) Z. *Pflanzenernaehr. Bodenkd.* **131**: 159-170.

Jyung WH, Camp ME (1976). The effect of Zn on the formation of ribulose diphosphate carboxylase in *Phaseolus vulgaris*. *Physiol. Plant.* **36**: 350-355.

Jyung WH, Ermann A, Schelender KK, Scala J (1975). Zinc nutrition and starch metabolism in *Phaseolus vulgaris*. *Plant Physiol.* **55**: 414-420.

Kaiser WH, Weiner H, Kandlbinder A, Tsai C-B, Roctul P, Sonoda M, Planchet E (2002). Modulation of nitrate reductase: Some new insights, an unusual case and a potentially important side reaction. *J. Exp Bot.* **53**: 875-882.

Kampfenkel K, Kushnir S, Babiychuk E, Inzé D, Van Montagu M (1995). Molecular characterization of a putative *Arabidopsis thaliana* copper transporter and its yeast homologue. *J. Biol. Chem.* **270**: 28479-28486.

Kannan S, Ramani S (1978). Studies on molybdenum absorption and transport in bean and rice. *Plant Physiol.* **62**: 179-181.

Kaplan D, Lips SH (1984). A comparative study of nitrate reduction and oxidation of glycolate. *Israel J. Bot.* **33**: 1-11.

Kastori R, Petrovic N (1989). Effect of boron on nitrate reductase activity in young sunflower plants. *J. Plant Nutr.* **12**: 621-632.

Kastori R, Plesnicar M, Pankoic D, Sakae Z (1995). Photosynthesis, chlorophyll fluorescence and soluble carbohydrates in sunflower leaves as affected by boron deficiency. *J. Plant Nutr.* **18**: 1751-1763.

Katoh, S (1977). Plastocyanin In: Encyclopedia of Plant Physiology Vol. *Photosynthesis I.* A Trebst, M Avron (ed.) Springer-Verlag, Berlin-Heidelberg-New York. 247-252.

Katyal JC, Randhawa NS (1983). *Micronutrients.* Fertilizer and Plant Nutrition Bulletin No. 77. FAO Rome.

Katyal JC, Sharma BD (1980). A new technique of plant analysis to resolve iron chlorosis, *Plant Soil.* **56**: 103-119.

Kaur NP, Sadana US, Nayyar VK (1988). Effect of manganese deficiency stress on yield and reproductive physiology of durum wheat. In *Proc. Internatl. Symp. on Manganese in Soils and Plants*. MJ Webb, RO Nable, RD Graham, RJ Hannam (eds) Adelaide: 45-46.

Kaur P, Kaur S, Padmanabhan SY (1979). Effect of manganese and iron on incidence on brown spot disease of rice. *Indian Phytopath.* **32**: 287-288.

Kelley PM, Izawa S (1978). The role of chloride ion in photosystem II 1. Effects of chloride on photosystem II electron transport and hydroxylamine inhibition. *Biochem. Biophys. Acta.* **502**: 198-210.

Kerridge PC, Cook BG, Everett ML (1973). Application of molybdenum trioxide in the seed pallet for subtropical pasture legumes. *Trop. Grassl.* **7**: 229-232.

Kessler B, Monselise SP (1959). Ribonuclease as guide for the determination of zinc deficiency in orchard trees: In *Plant Analysis and Fertilizer Problems.* W. Reuther (ed.). *Amer. Inst. Biol. Sci.* Washington DC. 314-322.

Kessler E, Arthur W, Brugger JE (1957). The influence of manganese and phosphate on delayed light emission, fluorescence, photoreduction and photosynthesis in algae. *Arch. Biochem. Biophys.* **71**: 326-335.

Khan HR, Mc Donald GK, Rengel Z (1998). Chickpea genotypes differ in their sensitivity to Zn deficiency. *Plant Soil* **198**: 11-18.

Khurana N, Chatterjee C, Agarwala SC (1991). Effect of manganese deficiency on yield of lentil (*Lens culinaris*). *Indian J. Agric. Sci.* **61**: 395-399.

Khurana N, Chatterjee C, Sharma CP (1999).Impact of mangenese stress on physiology and seed quality of pea (*Pisum sativum*). *Indian J. Agric. Sci.* **69**: 332-335.

Kim HJ, Cate GG, Crain RC (1993). Potassium channels in *Samanea saman* protoplasts controlled by phytochrome and the biological clock. *Science* **260**: 960-962.

Kim J, Mayfield S (1997).Protein disulphide isomerase as a regulator of chloroplast translation activation. *Science.* **278**: 1954 -1957.

Kim YG, Cha J, Chandrasegaran S (1996). Hybrid restriction enzymes: Zinc finger fusions to *Fok* 1 cleavage domain. *Proc. Natl. Acad. Sci. USA.* **93**: 1156-1160.

Kirk GJ, Loneragan JF (1988). Functional boron requirement for leaf expansion and its use as a critical value for diagnosis of boron deficiency in soyabean. *Agron. J.* **80**: 758-762.

Kisker C, Schindelin H, Rees DC (1997). Molybdenum cofactor containing enzymes: structure and mechanism. *Annu. Rev. Biochem.* **66**: 233-267.

Kitagishi K, Obata H (1986). Effect of zinc deficiency on the nitrogen metabolism of meristematic tissues of rice plants with reference to protein synthesis. *Soil Sci. Plant Nutr.* **32**: 397-405.

Kitagishi K, Obata H, Kondo T (1987). Effect of zinc deficiency on 80s ribosome content of meristimetic tissue of rice plant. *Soil Sci. Plant Nutr.* **33**: 423-430.

Klug A, Rhodes D (1988). 'Zinc fingers': a novel protein motif for nucleic acid recognition. *Trends Biochem Sci.* 12: 464-469.

Kneen BE, La Rue TA, Welch RM, Weeden NF (1990). Pleotropic effects of *Brz.* A mutation in *Pisum sativum* (L.) cv. 'Sparkle' conditioning decreased nodulation and increased iron uptake and leaf necrosis. *Plant Physiol.* **93**: 717-722.

Knight AH, Crooke WM, Burridge JC (1973). Cation exchange capacity, chemical composition and the balance of carboxylic acids in the floral parts of various plant species. *Ann. Bot.* **37**: 159-166.

Kobayashi A, Sakamoto A, Kubo K, Rybka Z, Kanno Y, Takatsuji H (1998). Seven Zn–finger transcription factors are expressed sequentially during the development of anther in petunia. *Plant J.* **13**: 571-576.

Kobayashi M, Matoh T, Azuma J (1995). Structure and glyosyl composition of the boron-polysaccharide complex of radish roots. *Plant Cell Physiol.* **36**: S 139.

Kobayashi M, Matoh T, Azuma J (1996). Two chains of rhamno-galacturonan II are cross-linked by borate-diol ester bonds in higher plant cell walls. *Plant Physiol.* **110**: 1017-1020.

Kobayashi M, Ohno K, Matoh T (1997). Boron nutrition of cultured tobacco BY-2 cells, II. Characterisation of the boron-polysaccharide complex. *Plant Cell Physiol.* **38**: 676-683.

Kobayashi T, Nakanishi H, Takahashi M, Kawasaki S, Nishizawa NK, Mori S (2001). *In-vivo* evidence that *Ids 3* from *Hordeum vulgare* encodes a deoxygenase that converts 2- deoxymugineic acid to mugineic acid in transgeneic rice. *Planta.* **212**: 864-871.

Kobayashi M, Mutoh T, Matoh T (2004). Boron nutrition of cultured to-bacco By-2 cells. IV. Genes induced under low boron supply. *J. Exp. Bot.* **55**: 1441-1443.

Korshunova YO, Eide D, Clark WG, Guerinot ML, Pakrasi HB (1999). The IRT1 protein from *Arabidopsis thaliana* is a metal transporter with broad specifity. *Plant Mol. Biol.* **40**: 37-44.

Koshiba T, Saito E, Ono N, Yamamoto N, Sato M (1996). Purification and properties of flavin-and molybdenum containing aldehyde oxidase from coleoptile of maize. *Plant Physiol.* **110**: 781-789.

Kosegarten H, Hoffman B, Mengel K (2001). The paramount importance of nitrate in increasing apoplastic pH of young sunflower leaves to induce Fe deficiency chlorosis, and the re-greening effect brough about by acidic foliar sprays. *J. Plant Nutr. Soil Sci.* 164: 155-163.

Kramer D, Romheld V, Landsberg E-C, Marschner H (1980). Induction of transfer cell formation by iron deficiency in the root epidermis of *Helianthus annuus. Planta* **147**: 335-339.

Kriedemann PE, Anderson JE (1988). Growth and photosynthetic response to manganese and copper deficiencies in wheat (*Triticum aestivum*) and barley grass (*Hordeum glaucum* and *Hordeum leporinum*). *Aust. J. Plant Physiol.* **15**: 429-446.

Kriedemann PE, Graham RD, Wiskich JT (1985). Photosynthetic disfunc-tion and *in vivo* changes in chlorophyll a fluorescence from manganese deficient wheat leaves. *Aust. J. Agric. Res.* **36**: 157-169.

Krizek DT, Bennett JH, Brown JC, Zaharieva T, Nooris KH (1982). Photochemical reduction of iron I. Light reactions. *J. Plant Nutr.* 5: 323-333.

Kröniger W, Remenberg H, Tadvos MH, Polle A (1995). Purfication and properties of manganese superoxide dismutase from Norway spruce (*Picea abies* L. Karst), *Plant Cell Physiol.* **36**: 191-196.

Krüger C, Berkowitz O, Stephan UW, Hell R (2002). A metal-binding member of the late embryogenesis abundant protein family transport iron in the phloem of *Ricinus communis* L. *J.Biol.Chem.* **277**: 25062-25069.

Lagrimini LM (1991). Peroxidase, IAA oxidase and auxin metabolism in transformed tobacco plants. *Plant Physiol.* **96**: S-77.

Laity JH, Lee BM, Wright PE (2001). Zinc finger proteins: New insights into structural and functional diversity. *Curr. Opin. Struct. Biol.* **11**: 39-46.

Lakanen E, Erviö RA (1971). A comparison of eight extractants for the determination of plant available micronutrients in soils. *Acta Agron. Fenn.* **123**: 223-232.

Lalitha R, George R, Ramasarma T (1985). Mevalonate decarboxylation in lemongrass leaves. *Phytochemistry.* **24**: 2569-2571.

Lamb C, Dixon RA (1997). The oxidative burst in plant disease resistance. *Annu. Rev. Plant Physiol. Plant Mol. Biol.* **48**: 251-275.

Landsberg E-C (1981). Organic acid synthesis and release of hydrogen ions in response to Fe deficiency stress of mono- and dicotyledonous plant species. *J. Plant Nutr.* **3**: 579-591.

Landsberg E-C (1982). Transfer cell formation in the root epidermis: A prerequisite for Fe-efficiency ? *J. Plant Nutr.* **5**: 414-432.

Landsberg E-C (1986). Function of rhizodermal transfer cells in the Fe stress response mechanism of *Capsicum annuum* L. *Plant Physiol.* **82**: 511-517.

Landsberg E-C (1989). Proton efflux and transfer cell formation as response to Fe deficiency of soybean in nutrient solution culture. *Plant Soil* **114**: 53-61.

Lasat MM, Pence NS, Garvin DF, Ebbs SD, Kochian LV (2000). Molecular physiology of zinc transport in the zinc hyperaccumulator *Thlaspi caerulescens* .*J. Exp. Bot.* **51**: 71-79.

Lascano FR, Gomez LD, Casano LM, Trippi VS (1998). Changes in glutathione reductase activity and protein content in wheat leaves and chloroplasts exposed to photoxidative stress. *Plant Physiol. Biochem.* **36**: 321-329.

Läuchli A (2002). Functions of boron in higher plants: Recent advances and open questions. *Plant biol.* **4**: 190-192.

Laurenzi M, Tipping AJ, Marcus SE, Knox JP, Fedrico R, Angilini R, Mc Pherson MJ (2001). Analysis of the distribution of copper amine oxidase in cell walls of legume seedlings. *Planta.* **214**: 37-45.

Lawrence K, Bhalla P, Mishra PC (1995). Changes in NAD(P)H dependent redox activities in plasmalemma-enriched vesicles isolated from boron- and zinc-deficient chickpea roots. *J. Plant Physiol.* **146**: 652-657.

Layzell DB, Hunt S (1990). Oxygen and regulation of nitrogen fixation in legume nodules *Physiol. Plant.* **80**: 322-327.

Leaf GL (1953). Boron in relation to water balance in plants. *Proc. Iowa Acad. Sci.* **60**: 176-191.

Lee RB (1982). Selectivity and kinetics of iron uptake by barley plants following nutrient deficiency. *Ann. Bot.* **50**: 429-449.

Lee S, Aronoff S (1967). Boron in plants : A biochemical role. *Science.* **158**: 798-799.

Leece DR (1978). Distribution of physiological inactive zinc in maize growing on a black earth soil. *Aust. J. Agric. Res.* **29**: 749-758.

Leidi EO, Goméz M (1985). A role for manganese in the regulation of soybean nitrate reductase activity? *J. Plant Physiol.* **118**: 335-342.

Leidi EO, Gomez M, De La Guardia MD (1987a). Soybean genetic differences in response to Fe and Mn: Activity of metalloenzymes. In *Genetic Aspects of Plant Nutrition.* WH Gabelman, BC Loughman (eds). Maitinus Nijhoff Publishers, Dordrecht: 463-470.

Leidi EO, Gomez M, del Rio LA (1987b). Evaluation of biochemical indicators of Fe and Mn nutrition for soybean plant II. Superoxide dismutase, chlorophyll contents and photosystem II activity. *J. Plant Nutr.* **10**: 261-271.

Leseure AM, Proudhon D, Pesey H, Ragland M, Theil EC, Briat, JF (1991). Ferritin gene transcription is regulated by iron in soybean cell culture. *Proc. Natl. Acad. Sci. USA.* **88**: 8222-8226.

Lessani H, Marschner H (1978). Relation between salt tolerance and long-distance transport of sodium and chloride in various crop species. *Aust. J. Plant Physiol.* **5**: 27-37.

Lewis DH (1980a). Boron, lignification and the origin of vascular plants. *New Phytol.* **84**: 209-229.

Lewis DH (1980b). Are there inter-relations between the metabolic role of boron, synthesis of phenolic phytoelexins and the germination of pollen? *New Phytol.* **84**: 261-270.

Li C, Dannel F, Pfeffer H, Römheld V, Bangerth F (2001). Effect of boron starvation on boron compartmentation, and possibly hormone-mediated elongation growth and apical dominance in pea (*Pisum sativum*) plants. *Physiol. Plant.* **111**: 212-219.

Li YQ, Cai G, Mascatelli A, Cresti M (1997). Functional interaction among cytoskeleton, membranes and cell wall in the pollen tubes of flowering plants. *Int. Rev. Cytol.* **176**: 133-199.

Liang F, Cunningham KW, Harper JF, Sze H (1997). ECA1 complements yeast mutants defective in Ca^{2+} pumps and encodes an endoplasmic reticulum-type Ca^{2+} -ATPase in *Arabidopsis thaliana. Proc. Natl. Acad. Sci. USA.* **94**: 8579-8584.

Liang J, Karamanos RE (1993). DTPA extrachable Fe Mn, Cu and Zn In: *Soil Sampling and Methods of Analysis.* MR Carter (ed.) *Canadian Soc. Soil Sci.* Lewis Publishers: 87-90.

Licht S, Gerfen GJ, Stubbe, J (1996). Thiyl radicals in ribonucleotide reductase *Science* **271**: 477-487.

Lin CH, Stocking CR (1978) Influence of leaf age, light, dark, and iron deficiency on polyribosome levels in maize leaves. *Plant Cell Physiol.* **19**: 461-470.

Lin S-F, Baumer JS, Ivers D, Cianzio SR, Shoemaker RC (1998). Field and nutritient solution tests measure similar mechanisms controlling iron deficiency chlorosis in soybeans. *Crop Science* **38**: 254-259.

Lindsay CO, Rodrigues L, Pasternak CA (1989). Protection of cells against membrane damage by haemolytic agents: Divalent cations and protons act at the extracellular side of the plasma membrane. *Biophys. Acta.* **983**: 56-64.

Lindsay WI, Norvell WA (1978). Development of a DTPA soil test for zinc, iron, manganese, and copper. *Soil Sci. Soc. Amer. Proc.* **42**: 421-428.

Ling H-Q, Bauer P, Bereczky Z., Keller B, Ganal M (2002). The tomato *fer* gene encoding a bHLH protein controls iron-uptake responses in roots. *Proc. Natl. Acad. Sci.* USA **99**: 13938-13943.

Ling HQ, Koch G, Baumlein H, Ganal MW (1999). Map based cloning of *Chloronerva*, a gene involved in iron uptake of higher plants encoding nicotianamine synthase. *Proc. Niatl. Acad. Sci.* USA. **96**: 7078-7103.

Ling H-Q, Pich A, Scholz G, Ganal MW (1996) Genetic analysis of two tomato mutants affected in the regulation of iron metabolism. *Mol. Gen. Genet.* **252**: 87-92.

Lippard SJ (1999). Free copper ions in the cell. *Science* 284: 748-749.

Liu Z, Shi W, Fan X (1990). The rhizosphere effects of phosphorus and iron in soils. In *Transactions 14th Internatl. Cong. Soil Sci.* (M. Koshino, ed.) *Intern. Soc. Soil Sci.* Tokyo. 147-152.

Llamas A, Kalakontskii KL, Fernandez E (2000). Molybdenum cofactor amounts in *Chlamydomonas reinhardtii* depend on the *Nit5* gene function related to molybdate transport. *Plant Cell Environ.* **23**: 1247-1255.

Lloyd A, Christopher LP, Carroll D, Drews GN (2005). Targeted mutagenesis using zinc–finger nucleases in *Arabidopsis. Proc. Natl. Acad. Sci.* USA. **102**: 2232-2237.

Lobreaux S, Massenet O, Briat J-F (1992). Iron induces ferretin synthesis in maize plantlets. *Plant Mol. Biol.* **19**: 563-575.

Lohaus G, Hussmann M, Pennewiss K, Schneider H, Zhu J-J, Sattelmacher B (2000). Solute balance of a maize (*Zea mays*) source leaf as affected by salt treatment with special emphasis on phloem retranslocation and ion leaching. *J. Exp. Bot.* **51**: 1721-1732.

Lohins MP (1937). Plant development in the absence of boron. Overgedrukt uit Mededeelingen van de Landbourwhooge-school. **41**: 3-36.

Loneragan JF, Delhaize E, Webb J (1982). Enzymic diagnosis of copper deficiency in subterranian clover I. Relationship of ascorbate oxidase activity in leaves to plant copper status. *Aust. J. Agric. Res.* **33**: 967-979.

Loneragan JF, Robson AD, Graham RD (eds) (1981) *Copper in Soils and Plants.* Academic Press, London.

Loneragan, JF (1981). Distribution and movement of copper in plants. In: *Copper in Soils and Plants.* JF Loneragan, AD Robson, RD Graham (eds) Academic Press, London. 165-188.

Longnecker N, Crosbie J, Davies F, Robson A (1996). Low seed manganese concentration and decreased emergence of *Lupinus angustifolius. Crop Sci.* **36**: 355-361.

Longnecker NE, Graham RD, Card G (1991a). Effects of manganese deficiency on the pattern of tillering and development of barley (*Hordeum vulgare* cv. Galleon). *Field Crops Res.* **28**: 85-102.

Longnecker NE, Marcar NE, Graham RD (1991b). Increased manganese content of barley seed can increase grain yield in manganese-deficient conditions. *Aust. J. Agric. Res.* **42**: 1065-1074.

Longnecker NE, Welch RM (1990). Accumulation of apoplastic iron in plant roots. A factor in the resistance of soybean to iron deficiency induced chlorosis? *Plant Physiol.* **92**: 17-22.

Loomis WD, Durst RW (1991b). Boron and cell walls. *Curr. Top. in Plant Biochem. Physiol.* **10**: 149-178.

Loomis WD, Durst RW (1992). Chemistry and biology of boron. *Bio Factors* **3**: 229-239.

Loomis WE, Shull CA (1937). *Methods in Plant Physiology* McGraw Hill, New York.

Lopez-Millan AF, Morales F, Abadia A, Abadia J (2001). Changes induced by Fe deficiency and Fe resupply in the organic acid metabolism of sugarbeet (*Beta vulgaris*) leaves. *Physiol. Plant.* **112**: 31-38

Lorenzen I, Aberle T, Plieth C (2004). Salt stress-induced chloride flux: A study using transgenic *Aarabidopsis* expressing a fluorescent anion probe. *Plant J.* 38: 539-544.

Lucas RE, Knezek BD (1972). Climatic and soil conditions promoting micronutrient deficiencies in plants. In: "Micronutrients in Agriculture. JJ Mortvedt, PM Giordano, WL Lindsay (eds). *Soil Sci. Soc. Amer.* Madison WI. 265-288

Lucaszewski KM, Blevins DG (1996). Root growth inhibition in boron deficient or aluminium-stressed squash plants may be a result of impaired ascorbate metabolism *Plant Physiol.* **112**: 1-6.

Lucaszewski KM, Blevins DG, Randall DD (1992). Asparagine and boric acid cause allantoate accumulation in soybean leaves by inhibiting manganese-dependent allantoate amidohydrolase. *Plant Physiol.* **99**: 1670-1976.

Lynch J (1998). The role of nutrient efficient crops in modern agriculture. In *Nutrient Use in Crop Production.* Z Rengel (ed.) Howarth Press Binghamton, NY. 241-264.

Ma JF, Kusano G, Kimura S, Nomoto K (1993). Specific recognition of mugineic acid-ferric complex by barley roots. *Phytochemistry* **34**: 599-603.

Ma JF, Nomoto K (1996). Effective regulation of iron acquisition in graminaceous plants. The role of mugineic acids as phytosiderophores. *Physiol. Plant* **97**: 609-617.

Ma JF, Shinada T, Matsuda C, Nomoto K (1995). Biosynthesis of phytosiderophores, mugineic acids, associated with methionine cycling. *J. Biol. Chem.* **270**: 16549-16554.

Maas FM, Van de Wetering DAM, van Beusichem ML, Bienfait HF (1988). Characterization of phloem iron and its possible role in the Fe-efficiency reactions. *Plant Physiol.* **87**: 167-171.

Mac Kay JP, Crossley M (1998). Zinc fingers are sticking together. *Trends Biochem. Sci.* **23**: 1-4.

Machold O (1968). Einfluss der Ernährungsbedingungen auf' den Zustand des Eisens in den Blättern, den Chlorophyllgehalt und die Katalase-sowie Peroxydase-aktivität. *Flora* (Jena) Abt. **A159**: 1-25.

Maksymiec W (1997). Effect of copper on cellular processes in higher plants. *Photosynthetica.* **34**: 321-342.

Malan C, Greyling MM, Gressel J (1990). Correlation between CuZn super-oxide dismutase and glutathione reductase, and environmental and xenobiotic stress tolerance in maize inbreds. *Plant Sci.* **69**: 157-166.

Manthey JJ, Tissarat B, Crowley DE (1996). Root responses of sterile-grown onion plants to iron deficiency. *J. Plant Nutr.* **19**: 145-161.

Marcar NE, Graham RD (1987). Genotypic variation for manganese efficiency in wheat. *J. Plant Nutr.* **10**: 2049-2055.

Marentes E, Shelp BJ, Vanderpool RA, Spiers GA (1997). Retranslocation of boron in broccoli and lupin during early reproductive growth. *Physiol. Plant.* **100**: 389-399.

Maret W, Jacob C, Vallee BL, Fischer EH (1999). Inhibitory sites in enzymes: Zinc removal and reactivation by thionein. *Proc. Natl. Acad. Sci.* USA **96**: 1936-1940.

Marschner H (1993). Zinc uptake from soils. In *Zinc in Soils and Plants.* AD Robson (ed.). Kluwer Academic Publishers, Dordrecht. 59-77

Marschner H (1995). *Mineral Nutrition of Higher Plants.* Academic Press, London.

Marschner H, Cakmak I (1989). High light intensity enhances chlorosis and necrosis in leaves of zinc, potassium and magnesium deficient bean (*Phaseolus vulgaris*) plants. *J. Plant Physiol.* **134**: 308-315.

Marschner H, Kirkby EA, Cakmak I (1996). Effect of mineral nutritional status on shoot-root partitioning of photoassimilates and cycling of mineral nutrients. *J. Exptl. Bot.* **47**: 1255-1263.

Marschner H, Römheld V (1994). Strategies of plants for acquisition of iron. *Plant Soil* **165**: 261-274.

Marschner H, Romheld V, Kissel M (1986). Different strategies in higher plants in mobilization and uptake of iron. *J. Plant Nutr.* **9**: 695-713.

Marsh HV Jr, Evans HJ, Matrone C (1963). Investigations on the role of iron deficiency on chlorophyll and haem content and the activities of certain enzymes in leaves. *Plant Physiol.* **38**: 632-637.

Martens DC, Lindsay WL (1990). Testing soils for copper, iron, manganese and zinc. In *Soil Testing and Plant Analysis.* RL Westerman (ed.) *Soil. Sci. Soc. Amer. Book Ser 3.* Madison, WI. 229-263.

Martens DC, Westermann DT (1991). Fertilizer application for correcting micronutrient deficiencies. In: *Micronutrients in Agriculture* 2nd ed. JJ Mortvedt, FR Cox, LM Schuman, RM Welch (eds) Soil Sci. Soc. Amer. Book Series Madison WI. 549-592.

Martin-Prìvel P, Gagnard J, Gautier P (1987). *Plant Analysis as a Guide to Nutrient Requirement of Temperate and Tropical Crops.* Lavoiser, New York, Paris.

Marziah M, Lam CA (1987). Polyphenol oxidase from soybeans (*Glycine max* cv. Palmetto) and its response to copper and other micronutrients. *J. Plant Nutr.* **10**: 2089-2094.

Masalha J, Cosegarten H, Elmaci O, Mengel K, (2000). The essential role of microbial activity for iron acquisition in maize and sunflower. *Biol. Fertil. Soil.* **30**: 433-439.

Mateo P, Bonilla I, Fernandez-Valiente E, Sanchez-Mase OE (1986). Essentiality of boron for dinitrogen fixation in *Anabaena* sp. PCC 7119. *Plant Physiol.* **81**: 430-433.

Matoh T, Ishigaki K, Ohno K, Azuma J (1993). Isolation and characterization of a boron-polysaccharide complex from radish roots. *Plant Cell Physiol.* **34**: 639-642.

Matoh T, Ishigaki K,Mizutani M, Matsunaga W, Takabe K (1992). Boron nutrition of cultured tobacco BY-2 cells. 1, Requirement for and intracellular localization of boron and selection of cells that tolerate low levels of boron. *Plant Cell Physiol.* **33**: 1135-1141.

Matoh T, Kawaguchi S, Kobayashi M (1996). Ubiquity of a borate-rhamnogalacturonan II complex in the cell walls of higher plants. *Plant Cell Physiol.* **37**: 636-640.

Matoh T, Kobayashi M (1998). Boron and calcium, essential inorganic constituents of pectic polysaccharides in higher plant cell walls. *J. Plant Res.* **111**: 179-190.

Maudinas B, Bucholtz ML, Papostephanon C, Katiyar SS, Briedis AV,Porter JW (1977). The partial purification and properties of a phytoene synthesizing enzyme system, *Arch. Biochem. Biophys.* **180**: 354-362.

Mauk MR, Kishi K, Gold MH, Mauk AG (1998). pH-linked binding of Mn (II) to manganese peroxidase. *Biochemistry* **37**: 6767-6771.

Maxwell DP, Wong Y, McIntosh L (1999). The alternative oxidase lowers mitochondrial reactive oxygen production in plant cells. *Proc. Natl. Acad. Sci. USA* **96**: 8271-8276.

Mc Cay-Buis TS, Huber DM, Graham RD, Phillips JD, Miskin KE (1995). Manganese seed content and take-all of cereals. *J. Plant. Nutr.* **18**: 1711-1721.

Mc Kersie BD, Chen Y, de Beus M, Bowley SR, Bowler C, Inze D, D'Halluin, Botterman J (1993). Superoxide dismutase enhances tolerance of freezing stress in transgenic alfalfa (*Medicago sativa* L.) *Plant Physiol.* **103**: 1155-1163.

Mehdy MC (1994). Active oxygen species in plant defense against pathogen. *Plant Physiol.* **105**: 467-472.

Mehlich A (1984). Mehlich 3 soil extractant: A modification of Mehlich 2 extractant. *Commun. Soil Sci. Plant Anal.* **15**: 1409-1416.

Mehlorn H, Wenzel A (1996). Manganese deficiency enhances ozone toxicity in bush bean (*Phaseolus vulgaris* L. cv. Saxa). *J. Plant Physiol.* **148**: 155-159.

Mehrotra SC, Jain P. (1996). Evaluation of some biochemical parameters used for diagnosis of iron chlorosis in vegetable plants. *Indian J. Agric. Sci.* **9**: 47-51.

Mehrotra SC, Gupta P (1990). Reduction of iron by leaf extracts and its significance for the assay of Fe (II) iron in plants. *Plant Physiol.* 93: 1017-1020.

Mehrotra SC, Sharma CP, Agarwala SC (1985). A search for extractants to evaluate the iron status of plants. *Soil Sci. Plant Nutr.* **31**: 155-162.

Melsted SW, Matto HL, Peck TR (1969). Critical plant nutrient composition values useful in interpreting plant analysis data. *Agron. J.* **61**: 17-20.

Mendel RR (1997). Molybdenum cofactor of higher plants : Biosynthesis and molecular biology. *Planta* **203**: 399-405.

Mendel RR, Hansch (2002). Molybdoenzymes and molybdenum cofactor in plants. *J, Exp. Bot.* **53**: 1689-1698.

Mendel RR, Müller AJ (1976). A common genetic determinant of xanthine dehydrogenase and nitrate reductase in *Nicotiana tabacum*. *Biochem. Physiol. Pflanzen.* **170**: 538-541.

Mendel RR, Schwarz G (1999). Molybdoenzymes and molybdenum cofactors in plants. *Crit. Rev. Plant Sci.* **18**: 33-69.

Mengel K (1994). Iron availability in plant tissues—iron chlorosis on calcareous soils. *Plant Soil* 165: 275-283.

Mengel K, Breinniger MT, Bübl W (1984). Bicarbonate the most important factor inducing iron chlorosis in vine grapes on calcareous soil. *Plant Soil* **81**: 333-344.

Mengel K, Geurtzen G (1988). Relationship between iron chlorosis and alkalinity in *Zea mays*. *Physiol. Plant* **72**: 460-465.

Mengel K, Kirkby EA (2001). *Principles of Plant Nutrition* 5th ed. Kluwer Academic Publishers, Dordrecht.

Merchant S, Dreyfuss BW (1998). Post translational assembly of photosynthetic metalloproteins. *Annu. Rev. Plant Physiol. Plant Mol. Biol*. **49**: 25-51.

Merkel D, Witt HH, Jungk A (1975). Effect of molybdenum on the cation–anion balance of tomato plants at different nitrogen nutrition. *Plant Soil*. **42**: 131-143.

Millenaar FF, Benschop JJ, Wagner AM, Lambers H (1996). The role of alternate oxidase in stabilizing the *in vitro* reduction state of the ubiquinone pool and the activation state of the alternate oxidase. *Plant Physiol*. **118**: 599-607.

Miller GW, Pushnik JC, Welkie GW (1984). Iron chlorosis, a worldwide problem, the relation to chlorophyll biosynthesis to iron. *J. Plant Nutr*. **7**: 1-22.

Miller GW, Shigematsu A, Welkie GW, Motoji N, Szlek M (1990). Patassium effect on iron stress in tomato II. The effects in root CO_2 fixation and organic acid fixation *J. Plant Nutr*. **13**: 1355-1370.

Miyake C, Asada K (1994). Ferredoxin-dependent photoreduction of the monodehydroascorbate radical in spinach thylakoids. *Plant Cell Physiol*. **35**: 539-549.

Miyake C, Cao W-H, Asada K (1993). Purification and molecular properties of the thylakoid – bound ascorbate peroxidase in spinach chloroplasts. *Plant Cell Physiol*. **34**: 881-889.

Mizuno N, Inazu O, Kamada K (1982). Characteristics of concentrations of copper, iron and carbohydrates in copper deficient wheat plants. In *Proceedings 9th Internatl. Plant Nutr. Colloquium* Warwick, England. A Scaife (ed.) Commonw. Agric. Bur. Farnham Royal, Bucks: 396-399.

Mohammad MJ, Najim H, Kheresat S (1998). Nitric acid and O-phenanthroline extractable iron for diagnosis of iron chlorosis in citrus lemon trees. *Commun. Soil. Sci. Plant. Anal*. **29**: 1035-1043.

Mondy NI, Munshi CB (1993). Effect of boron on enzymatic discoloration and phenolic and ascorbic acid content of potatoes. *J. Agric. Food Chem*. **41L** 554-556.

Montalbini P (1992). Inhibition of hypersensitive response by allopurinol applied to the host in the incompatible relationship between *Phaseolus vulgaris* and *Uromyces phaseoli*. *J. Phytopath*. **134**: 218-228.

Montalbini P (1998). Purification and some properties of xanthine dehydrogenase from wheat leaves. *Plant Sci*. **134**: 89-102.

Moog PR, Bruggemann W (1995). Iron reductase systems on the plant plasma membrane - a review. *Plant Soil* **165**: 241-260.

Moore jr PA, Patrick jr WH (1988). Effect of zinc deficiency on alcohol dehydrogenase activity and nutrient uptake in rice. *Agron. J*. **80**: 882-885.

Morab H, Koti RV, Chetti MB, Patil PV, Nalini AS (2003). Role of nutrients in inducing rust resistance in soybean. *Indian J. Plant Physiol.* **8**: 85-88

Moraghan JT (1980). Effect of soil temperature on response of flux to P and Zn fertilizers. *Soil Sci.* **129**: 290-296.

Moraghan JT (1991). Removal of endogenous iron, manganese and zinc during plant washing. *Commun. Soil Sci. Plant Anal.* **22**: 323-330.

Morales F, Abadia A, Abadia J (1990). Characterization of the xanthophyll cycle and other photosynthetic pigment changes induced by iron deficiency in sugarbeet (*Beta vulgaris* L.). *Plant Physiol.* **94**: 607-613.

Morales F, Susin S, Abadia A, Cauera M, Abadia J (1992). Photosynthetic characteristics of iron chlorotic pear (*Pynus communis* L.). *J. Plant Nutr.* **15**: 1783-1790.

Moreau S, Thomson RM, Kaiser BN, Trevaskis B, Guerinot ML, Udvardi MK, Puppo A, Day DA (2002). *Gm ZIP1* encodes a symbiosis-specific zinc transporter in soybean. *J. Biol. Chem.* **277**: 4738-4746.

Morgan PW, Taylor DM, Joham HE (1976). Manipulation of IAA oxidase activity and auxin deficiency symptoms in intact cotton plants with manganese nutrition. *Physiol. Plant.* **37**: 149-156.

Morghan JT, Grafton K (1999). Seed-Zn concentration and zinc efficiency trait in navy bean. *Soil Sci. Soc. Amer. J.* **63**: 918-922.

Mori S, Kishi-Nishizawa N, Fujigaki J (1990). Identification of rye chromosome 5R as a carrier of the gene for mugineic acid synthase and 3-hydroxymugineic acid synthase using wheat rye addition lines. *Jpn. J. Genet.* **65**: 343-352.

Mori S, Nishizawa N (1987). Methionine as a dominant precursor of phytosiderophores in graminaceae plants. *Plant Cell Physiol.* **28**: 1081-1092.

Mori S, Nishizawa N, Hayashi H, Chino M, Yoshimura E, Ishihara J (1991). Why are young rice plants highly susceptible to iron deficiency? *Plant Soil* **130**: 143-156.

Mortvedt JJ (1991). Correcting iron deficiencies in annual and perennial plants. Present technologies and future prospects. *Plant Soil* **130**: 273-279.

Mortvedt JJ (1997). Sources and methods for molybdenum fertilization of crops. In: *Molybdenum in Agriculture.* UC Gupta (ed.) Cambridge University Press, New York. 171-181.

Mortvedt JJ (ed.) (1991). *Micronutrients in Agriculture* 2nd ed. *Soil. Sci. Soc. Amer.* Book Ser. 4 Madison, WI.

Mortvedt JJ, Anderson OE (1982). *Forage legumes: Diagnosis and Correction of Molybdenum and Manganese Problems.* Univ. Georgia Southern Coop. Ser. Bull. 278.

Moseley JL, Allinger T, Herzog S, Hoerth P, Wehinger E, Merchant S, Hippler M (2002). Adaptation to Fe-deficiency requires remodeling of the photosynthetic apparatus. *EMBO J.* **21**: 6709-6720.

Moudinas B, Bucholtz ML, Papastephanou C, Katiyar SS, Briedis AV, Porter JW (1977). The partial purfication and properties of phytoene synthesizing enzyme system *Arch. Biochem. Biophys.* **180**: 354-362.

Moussavi-Nik M, Rengel Z, Hollamby GJ, Ascher JS (1997). Seed manganese (Mn) content is more important than Mn fertilization for wheat growth under Mn deficient conditions. In *Plant Nutrition For Sustaniable Food Production and Environment.* T Ando, K Fujita, T Mae, H Matsumoto, S Mori, J Sekiya (eds) Kluwer Academic Publishers, Dordrecht. 267-268.

Moya JL, Primo-Millo E, Talon M (1999). Morphological factors determining salt tolerance in citrus seedlings: the shoot to root ratio modulates passive root uptake of chloride ions and their accumulation in leaves. *Plant Cell Environ.* **22**: 1425.

Mozafar A (1993). Role of boron in seed production. In *Boron and its Role in Crop Production.* UC Gupta (ed.) CRC Press Boca. Raton FL.

Mulder, EG (1939). On the use of microorganisms in measuring the deficiency of copper, magnesium and molybdenum. *Antonie van Leuwenhoek* **6**: 99-109.

Mullins GL, Sommers LE (1986). Cadmium and zinc influx characteristics by intact corn (*Zea mays* L) seedlings. *Plant Soil.* **96**: 153-164.

Munsen RD, Nelson WL (1990). Principles and practices in plant analysis. In *Soil Testing and Plant Analysis.* RL Westerman (ed.) *Soil Sci. Soc. Amer. Book Ser 3.* Madison WI. 357-38.

Murphy LS, Walsh LM (1991). Correction of micronutrient deficiencies with fertilizers. In *Micronutrients in Agriculture.* JJ Morvedt, FR Cox, LM Shuman, RM Welch (eds) *Soil Sci. Soc. Am.* Madison WI. 347-387.

Nable RO, Banuelos GS, Paull JG (1997). Boron toxicity. *Plant Soil.* **193**: 181-198.

Nable RO, Bar-Akiva A, Loneragan JF (1984). Functional manganese requirement and its use as a critical value for diagnosis of manganese deficiency in subterranean clover (*Trifolium subterraneum* L. cv Seaton Park). *Ann. Bot.* **54**: 39-49.

Nable RO, Cartwright B, Lance RCM (1990). Genotypic differences in boron accumulation in barley: relative susceptibilies to boron deficiency and toxicity. In *Genetic Aspects of Plant Mineral Nutrition.* N El-Bassam, M Dambroth, BC Loughman (eds). Kluwer Academic Publishers, Dordrecht. 243-251.

Nakagawa H, Jiang CJ, Sakakibara H, Kojima M, Honda I, Ajisaka H, Nishijima T, Koshioka M, Homma J, Mander LN, Takatsuji H (2005). Overexpression of a petunia zinc finger gene alters cytokinin metabolism and plant forms. *Plant J.* **41**: 512-523.

Nakagawara H, Ferrario S, Angenent GC, Kabayashi A, Takatsuji H (2004). The petunia ortholog of *Arabidopsis SUPERMAN* plays a distinct role in floral organ morphogenesis. *Plant Cell.* **16**: 920-932.

Nakagawara H, Jiang C-J, Sakakibara H, Kojima M, Honda I, Ajisaka H, Nishijama T, Koshioka M, Homma T, Mander LN, Takatsuji H (2005). Overexpression of a petunia zinc finger gene alters cytokinin metabolism and plant forms. *Plant J.* **41**: 512-523.

Nakanishi H, Yamaguchi H, Sasakuma T, Nishiizawa N-K, Mori S (2000). Two dioxygenase genes, *Ids3* and *Ids2* from *Hordeum vulgare* are involved in the biosynthesis of mugineic acid family phytosiderophores. *Plant Mol. Biol.* **44**: 199-207.

Nambiar EKS (1976a). Uptake of Zn^{65} from dry soil by plants. *Plant Soil* **44**: 267-271.

Nambiar EKS (1976b). The uptake of Zn^{65} by roots in relation to soil water content and root growth. *Aust. J. Soil Sci.* **14**: 67-74.

Nason A, Lee K-Y, Pan S-S, Ketchum PA, Lamberti A, De Vries J (1971). In vitro formation of assimilatory reduced nicotinamide adenine dinucleotide phosphate nitrate reducate from a *Neurospora* mutant and a component of molybenum-enzyme. *Proc. Natl. Acad. Sci. USA.* **68**: 3242-3246.

Nautiyal N, Chatterjee C (1999). Role of copper in improving the seed quality of sunflower (*Helianthus annuus*). *Indian J. Agric. Sci.* **69**: 210-213.

Nautiyal N, Chatterjee C, Sharma CP (1999). Copper stress affects grain filling in rice. *Commun. Soil Sci. Plant Anal.* **30**: 1625-1632.

Nautiyal N, Singh S, Chatterjee C (2005). Seed reserves of chickpea in relation to molybdenum supply. *J. Sci. Food Agric.* **85**: 860-864.

Navarre DA, Wendehenne D, Durner J, Noad R, Klessig DF (2000). Nitric oxide modulates the activity of tobacco aconitase. *Plant Physiol.* **122**: 573-582.

Nayyar VK, Takkar PN (1980). Evaluation of various zinc sources for rice grown on alkali soils. *Z. Pflanzenernahr. Bodenk.* **143**: 489-493.

Negishi T, Nakanishi H, Yazaki J, Kishimoto N, Fujii F, et al. (2002). cDNA microarray analysis of gene expressions during Fe-deficiency stress in barley suggests that polar transport of vesicles is implicated in phytosiderophore secretion in Fe-deficient barely roots. *Plant J.* **30**: 83-94.

Neill SJ, Desikan R, Hancock JT (2003). Nitric oxide signalling in plants. *New Phytol.* **159**: 11-35.

Ness, PJ, Woolhouse HW (1980). RNA synthesis in *Phaseolus* chloroplasts. I. Ribonucleic acid and senescing leaves *J. Exp. Bot.* **31**: 223-233.

Nguyen J (1986). Plant xanthine dehydrogenase : Its distribution, properties and function. *Physiol. Veg.* **24**: 263-281.

Nguyen J, Feierabend J (1978). Some properties and subcellular localization of xanthine dehydrogenase in pea leaves. *Plant Sci. Lett.* **13**: 125-132.

Nicholas DJD, Fielding AH (1951). The use of *Aspergillus niger* (M) for determination of manganese, zinc, copper and molybdenum available in soil to crop plants. *J. Hort. Sci.* **26**: 125-147.

Nielson GG, Jensen A (eds) (1999). *Plant Nutrition — Molecular Biology and Genetics.* Kluwer Academic Publishers, Boston.

Nishio JN, Taylor SE, Terry N (1985). Changes in thylakoid galactolipids and proteins during iron nutrition-mediated chloroplast development. *Plant Physiol.* **77**: 705-711.

Noguchi K, Dannel F, Pfeffer H, Römheld V, Hayashi H, Fujiwara T (2000). Defect in root-shoot translocation of boron in *Arabidopsis thaliana* mutant *bor 1-1. Plant Physiol.* **156**: 751-755.

Noguchi K, Yasumori M, Imai T, Naito S, Matsunaga T, Oda H, Hayashi H, Chino.M, Fujiwara T (1997). *bor1-1* an *Arabidiopsis thaliana* mutant that requires a high level of boron. *Plant Physiol.* **115**: 901-906.

Normanly J, Slovin JP, Cohen JD (1995). Rethinking auxin biosynthesis and metabolism. *Plant Physiol.* **107**: 323-329.

Norvell WA, Welch RM (1993). Growth and nutrient uptake by barley (*Hordeum vulgare* L cv Herta): Studies using an N- (2 hydroxyethyl) ethylene dinitrilotriacetic acid buffered nutrient solution technique I. Zinc ion requirements *Plant Physiol.* **101**: 619-625.

O'Halloran TV, Culotta VC (2000). Metallo-chaperones, an intracellular shuttle service for metal ions. *J. Biol. Chem.* **275**: 25057-25060.

O'Hara GW, Dilworth MJ, Boonkerd N, Parkpian P (1988). Iron deficiency specially limits nodule development in peanut inoculated with *Bradyrhizobium* sp. *New Phytol.* **108**: 51-57.

O'Leary MH (1982) Phosphoenol pyruvate carboxylase: An enzymologists view. *Annu. Rev. Plant Physiol.* **33**: 297-315.

O'Neill MA, Eberhard S, Albersheim P, Darvill AG (2001). Requirement of borate cross-linking of cell wall rhamnogalacturonan II for *Arabidopsis* growth. *Science* **294**: 846-849.

O'Neill MA, Ishii T, Albusheim P, Darvill AG (2004). Rhamnogalacturonan II: Structure and function of a borate cross–linked cell wall pectic polysaccharide. *Annu. Rev. Plant Biol.* **55**: 109-139.

O'Neill MA, Warrenfeltz D, Kates K, Pellerin P, Doco T, Darvill AG, Albersheim P (1996). Rhamnogalacturonan-II, a pectic polysaccharide in the walls of growing plant cell, forms a dimer that is covalently cross-linked by a borate ester. *In vitro* conditions for the formation and hydrolysis of the dimer. *J. Biol. Chem.* **271**: 22923-22930.

O'Sullivan (1970). Aldolase activity in plants as an indicator of zinc deficiency. *J. Sci. Food Agric.* **21**: 607-609.

Obata H, Kawamura S, Shimoyara A, Senoo K, Tanaka A (2001). Free radical injury in rice leaf under Zn deficiency. *Soil Sci. Plant Nutr.* **47**: 205-211.

Obata H, Kawamura S, Senoo K, Tanaka A (1999). Changes in the level of protein and activity of Cu/Zn-SOD in zinc deficient rice plant (*Oryza sativa*). *Soil Sci. Plant Nutr.* **45**: 891-896.

228

Obata H, Umebayashi M (1988). Effect of zinc deficiency on protein synthesis in cultured tobacco plant cells. *Soil Sci. Plant Nutr.* **34**: 351-357.

Oberg G (1998). Chloride and organic chlorine in soil. *Acta Hydrochim. Hydrobiol.* **26**: 137-144.

Ocumpaugh WR, Rahmes JN, Bryn DW, Wei LC (1992). Field and greenhouse screening of oat seedlings for iron-nutrition efficiency. *J. Plant Nutr.* **15**: 1715-1725.

Oertli JJ, Grgurevic E (1975). Effect of pH on the absorption of boron by excised barley roots. *Agron. J.* **67**:278-280.

Oertli JJ (1993). The mobility of boron in plants. *Plant Soil* **155/156**: 301-304.

Oertli JJ, Richardson WF (1970). The mechanism of boron immobility is plants. *Physiol. Plant.* **23**: 108-116.

of ryegrass (*Lolium perenne* L. cv. Appolo). *J. Expt. Bot.* **51**: 945-953.

Ogawa K, Kanematsu S, Takabe K, Asada K (1995). Attachment of Cu-Zn superoxide dismutase to thylakoid membranes at the site of superoxide generation (PS I) in spinach chloroplasts: Detection by immuno-gold labeling after rapid freezing and substitution method. *Plant Cell Physiol.* **36**: 565-573.

Ohki K (1976). Effect of zinc nutrition on photosynthesis and carbonic anhydrase activity in cotton. *Physiol. Plant.* **38**: 300-304.

Okumura N, Nishizawa N-K, Umehara Y, Ohata T, Nakanishi H, et al. (1994). A dioxygenase gene (Ids2) expressed under iron deficiency conditions in the roots of *Hordeum vulgare*. *Plant Mol. Biol.* **25**: 705-719.

Olsen RA, Bennett JH, Blume D (1982). Reduction of Fe^{3+} as it relates to Fe chlorosis. *J. Plant Nutr.* **5**: 433-445.

Ono T (2001). Metallo-radical hypothesis for photoassembly of (Mn_4)–cluster of photosynthetic oxygen evolving complex. *Biochim. Biophys. Acta* **1503**: 40-51.

Ono T, Onone Y (1991). A possible role of redox-active histidine in the photoligation of manganese into photosynthetic O_2-evolving enzyme. *Biochemistry* **30**: 6183-6188.

Ordiz MI, Barbas III CF, Beachy RN (2002). Regulation of transgene expression in plants with polydactyl zinc finger transcription factors. *Proc. Natl. Acad. Sci. USA.* **99**: 13290-13295.

Orozeo-Cardinas M, Narvaez-Vasquez J, Ryan C (2001). Hydrogen peroxide acts as a second messenger for the induction of defense genes in tomato plants in response to wounding, systemin and methyl jasmonate. *Plant Cell.* **13**: 179-191.

Oserkowsky J (1933). Quantitative relation between chlorophyll and iron in green and chlorotic pear leaves. *Plant Physiol.* **8**: 449-468.

Otegui MS, Capp R, Staehelin LA (2002). Developing seeds of *Arabidopsis* store different minerals in two types of vacuoles and in the endoplasmic reticulum. *Plant Cell.* **14**: 1311-1327.

Palmer MJ, Dekock PC, Bacon JSD (1963). Changes in concentration of malic acid and citric acid, Ca and K in the leaves during the growth of normal and iron deficient mustard plants. *J. Biochem.* **86**: 484-493.

Pandey N and Sharma CP (1989). Zinc deficiency effect on photosynthesis and transpiration in safflower and its reversal on making up the deficiency. *Indian J. Expt. Biol.* **27**: 376-377.

Pandey N, Gupta M, Sharma CP (1995a). Ultrastructural changes in pollen grains of green gram subjected to copper deficiency. *Geophytology* **25**: 147-150.

Pandey N, Gupta M, Sharma CP (1995b). SEM studies on Zinc deficient pollen stigma of *Vicia faba*. *Phytomorphology* **45**: 169-173.

Pandey N, Gupta M, Sharma CP (2000). Zinc deficiency affects pollen structure and viability in green gram. *Geophytology* **28**: 31-34.

Pandey N, Nautiyal BD, Sharma CP (2002a). Cotton response to manganese stress and amelioration of stress. *J. Indian Bot. Soc.* **81**: 93-95.

Pandey N, Pathak GC, Sharma CP (2006). Zinc is critically required for pollen formation and fertilization in lentil. *J. Trace Elem. Med. Biol.* (in press)

Pandey N, Pathak GC, Singh AK, Sharma CP (2002b). Enzymic changes in response to zinc nutrition. *J. Plant Physiol.* **151**: 1151-1153.

Pandey N, Sharma CP (1996). Copper effect on photosynthesis and transpiration in safflower. *Indian J. Exp. Biol.* **34**: 821-822.

Pandey N, Sharma CP (1998). Safflower in response to varying levels of zinc supply. *J. Indian bot. Soc.* **77**: 11-34.

Pandey N, Sharma CP (2000). Carbonic anhydrase activity and stomatal morphology associated with zinc deficiency induced changes in faba bean. *Phytomorphology.* **50**: 261-265.

Pandey N, Sharma CP (2002a). Metabolic changes in cotton plants subjected to copper deficiency and recovery. *Indian J. Plant Physiol.* **7**: 31-34.

Pandey N, Sharma CP (2002b). Effect of heavy metals Co^{2+}, Ni^{2+}, Cd^{2+} on growth and metabolism of cabbage. *Plant Sci.* **163**: 753-758.

Pandey N, Sharma CP (2003). Chromium interference in iron nutrition and water relations of cabbage. *Exp. Environ. Bot.* **49**: 195-200.

Pandey N, Singh AK, Pathak GC, Sharma CP (2002c). Effect of zinc on antioxidant response in maize (*Zea mays* L.) leaves. *Indian J. Exp. Biol.* **40**: 954-956.

Parker DR (1993). Novel nutrient solutions for zinc nutrition research: Buffering free $zinc^{2+}$ with synthetic chelators and P with hydroxylamine. *Plant Soil.* **155/156**: 461-464.

Parker DR (1997). Responses of six crop species to solution $Zinc^{2+}$ activities buffered with HEDTA. *Soil Sci. Soc. Am. J.* **61**: 167-176.

Parker DR, Gardner EH (1981). The determination of hot-water soluble boron in some acid Oregon soils using a modified azomethane H procedure. *Commun. Soil Sci. Plant Anal.* **12**: 1311-1322.

Parker DR, Norvell WA (1999). Advances in solution culture methods for plant mineral nutrient research. *Adv. Agron.* **65**: 151-313.

Parker MB, Harris HB (1977). Yield and leaf nitrogen of nodulating and non-nodulating soybean as affected by nitrogen and molybdenum. *Agronomy J.* **69**: 551-554.

Parr AJ, Loughman BC (1983). Boron and membrane functions in plants. In *Metals and Micronutrients : Uptake and Utilization by Plants.* Phytochem. Soc. Eur. Symp. Ser. No. 21. DA Robb, WS Pierpoint (eds.) Academic Press, London. 87-107.

Pastori GM, Rio LA (1997). Natural senescence of pea leaves: an activated oxygen-mediated function for peroxisomes. *Plant Physiol.* **113**: 411-418.

Pate JS, Hocking PJ (1978). Phloem and xylem transport in the supply of minerals to a developing legume (*Lupinus albus* L.) fruit. *Ann. Bot.* **42**: 911-921.

Pateman JA, Cove DJ, Rever BM, Roberts DB (1964). A common cofactor for nitrate reductase and xanthine dehydrogenase which also regulates the synthesis of nitrate reductase. *Nature* **201**: 58-60.

Paul EA, Clark FE (1996). *Soil Microbiology and Biochemistry.* Academic Press, London.

Pearson JN, Rengel Z (1994). Distribution and remobilization of ^{65}Zn and ^{54}Mn during grain development of wheat. *J. Exp. Bot.* **45**: 1829-1835.

Pearson JN, Rengel Z (1995a). Uptake and distribution of Zn and Mn in wheat grown at sufficient and deficient levels of ^{65}Zn and ^{54}Mn: I During vegetative growth. *J. Exp. Bot.* **46**: 833-839.

Pearson JN, Rengel Z (1995b). Uptake and distribution of Zn and Mn in wheat grown at sufficient and deficient levels of ^{65}Zn and ^{54}Mn: II During grain development. *J. Exp. Bot.* **46**: 841-845.

Pearson JN, Rengel Z (1997). Mechanisms of plant resistance to nutrient deficiency stresses. In *Mechanisms of Environmental Stress resistance in Plants.* AS Basra, RK Basra (eds) Harwood Academic Publishers, Amsterdam. 213-240.

Pearson JN, Rengel Z, Jenner CF, Graham RD (1995). Transport of zinc and manganese to developing wheat grains. *Physiol. Plant* **95**: 449-455.

Pedler JF, Webb MJ, Buchhorn SC, Graham RD (1996). Manganese oxidizing ability of isolates of the take-all fungus correlates to virulance. *Biol. Fert. Soils.* **22**: 272-278.

Pellerin, P, Doco T, Vidal S, Williams P, Brillouet J-M, O'Neill MA (1996). Structural characterization of red wine rhamnogalacturonan II. *Carbohydr. Res.* **290**: 183-197.

Perl A,Perl-Treves R, Galili S, Aviv D, Shalgi E, Malkin S,Galun E(1993). Enhanced oxidative stress defence in transgenic potato expressing tomato Cu, Zn superoxide dismutase. *Theor. App. Gen.* **85**: 568-576.

Perumal A, Beattie JM (1966). Effect of different levels of copper on the activity of certain enzymes in leaves of apple. *Proc. Am. Soc. Hortic. Sci.* **88**: 41-47.

Perur NG, Smith RL, Wiebe HH (1961). Effect of iron chlorosis on protein fraction on corn leaf tissue. *Plant Physiol.* **36**: 736-739.

Petranyl P, Jendrisak, JJ, Burgess RR (1978) RNA Polymerase II from wheat germ contains tightly bound zinc. *Biochem. Biophys. Res. Commun.* **74**: 1031-1038.

Pfeffer H, Dannel F, Romheld V (1997). Compartmentation of boron in roots and its translocation to the shoot of sunflower as affected by short term changes in boron supply. In *Boron in Soils and Plants.* RW Bell, B Rerkasem (eds) Kluwer Academic Publishers. Dordrecht. 203-207.

Pfeffer H, Dannel F, Romheld V (1998). Are there connections between phenol metabolism, ascorbate metabolism and membrane integrity in leaves of boron deficient sunflower plants. *Physiol. Plant.* **104**: 479-485

Pfeffer H, Dannel F, Romheld V (1999a). Isolation of soluble boron complexes and their determination together with free boric acid in higher plants. *J. Plant Physiol.* 154: 283-288.

Pfeffer H, Dannel F, Romheld V (1999b). Are there two mechanisms for boron uptake in sunflower. *J. Plant Physiol.* **155**: 34-40.

Pfeffer H, Dannel F, Romheld V (2001). Boron compartmentation in roots of sunflower plants of different boron status : A study using the stable isotopes ^{10}B and ^{11}B adopting two independent approaches. *Physiol. Plant.* **113**: 346-351.

Pich A, Manteuffel R, Hillman S, Scholz G, Schmidt W (2001) Fe homeostasis in plant cells: does nicotianamine play multiple roles in the regulation of cytoplasmic Fe concentrations ? Planta **213**: 967-976.

Pich A, Scholz G, Seifert K (1991). Effect of nicotinamine on iron uptake and citrate accumulation in two genotypes of tomato, *Lycoperiscon esculentum* Mill. *J. Plant Physiol.* **137**: 323-326.

Pierzynski GM, Jacobs LW (1986). Extractability and plant availability of molybdenum from inorganic and sewage sludge sources. *J. Environ. Qual.* **15**: 323-326.

Pinton R, Cakmak I, Marschner H (1993). Effect of zinc deficiency on proton fluxes in plasma membrane vesicles isolated from bean roots. *J. Exp. Bot.* **44**: 623-630.

Pinton R, Cakmak I, Marschner H (1994). Zn deficiency enhanced NADPH dependent superoxide radical production in plasma membrane vesicles isolated from root of bean plants. *J. Exp. Bot.* **45**: 45-50.

Pirson A (1937). Ernährungs – und stoffwechsel physiologische Untusuchungen an *Frontalis und Chlorella.* Z. Bot. **31**: 193-267.

Pitman MG (1969). Stimulation of Cl⁻ uptake by low salt barley roots as a test of models of salt uptake. *Plant Physiol.* **44**: 1417-1427.

Pitman MG (1982). Transport across plant roots. *Quart. Rev. Biophys.* **15**: 481-554.

Plank DW, Gengenbach BG, Gronwald JW (2001). Effect of iron on activity of soybean multi-subunit acetyl-coenzyme A carboxylase. *Physiol. Plant.* **112**: 183-194.

Platt-Aloia KA, Thomson WW, Terry N (1983). Changes in plastid ultra-structure during iron nutrition mediated chloroplast development. *Protoplasma* **114**: 85-92.

Polar E (1970). The Distribution of Zn^{65} in the cotyledons of *Vicia faba* and its translocation during the growth and maturation of the plant. *Plant Soil.* **32**: 1-17.

Pollard AS, Parr AJ, Loughman BC (1977). Boron in relation to membrane function in higher plants. *J. Exp. Bot.* **28**: 831-841.

Polle A, Chakrabarti K, Chakrabarti S, Seifert F, Schramel P, Rennenberg H (1992). Antioxidant and manganese deficiency in needles of Norway spruce (*Picea abies* L.) trees. *Plant Physiol.* **99**: 1084-1089.

Possingham JV, Vesk M, Mercer FV (1964). The fine structure of leaf cells of manganese deficient spinach. *J. Ultrastruct. Res* **11**: 69-83.

Power PP, Woods WG (1997). The chemistry of boron and its speciation in plants. *Plant Soil.* **193**: 1-13.

Prescott AG, John P (1996): Dioxygenases: Molecular structure and role in plant metabolism. *Annu. Rev. Plant. Physiol. Plant Mol. Biol.* **47**: 245-271.

Price AH, Hendry GAF (1991). Iron-catalyzed oxygen radical formation and its possible contribution to drought in nine native grasses and three cereals. *Plant Cell Environ.* **14**: 477-484.

Price CA, Clark HE, Funkhouser HE (1972). Functions of micronutrients in plants. In *Micronutrients in Agriculture*. JJ Mortvedt, PM Giordano, WL Lindsay (eds) Soil Sci. Soc. Amer. Madison WI. 731-742. *Principles.* HMSO, London.

Purcell LC, King CA, Ball RA (2000). Soybean cultivar differences in ureides and relationship to drought-tolerant nitrogen fixation and man-ganese nutrition. *Crop Sci.* **40**: 1062-1070.

Purvis AC (1997). Role of the alternative oxidase in limiting superoxide production by plant mitochondria. *Physiol. Plant.* **100**: 165-170.

Pushnik JC, Miller GW (1989). Iron regulation of chloroplast photosyn-thetic function: mediation of PS I development. *J. Plant Nutr.* **12**: 407-421.

Qadar A (2002). Selecting rice genotypes tolerant to zinc deficiency and sodicity stresses I. Differences in zinc, iron, manganese, copper, phos-phorus concentrations, and phosphorus/zinc ratio in their leaves. *J. Plant Nutr.* **25**: 457-473.

Quilez R, Abadia A, Abadia J (1992). Characteristics of thylakoids and photosystem II membrane preparations from iron deficient and iron suf-ficient sugarbeet (*Beta vulgaris* L.). *J. Plant Nutr.* **15**: 1809-1819.

Quinland SC, Miller GW (1982). The effect of Zn deficiency on the aldolase activity in the leaves of oats and clover. *Biochem J.* **53**: 457-460.

Rabotti G, de Nisi P, Zocchi G. (1995). Metabolic implications in the biochemical responses to iron deficiency in cucumber (*Cucumis sativus* L.) roots. *Plant Physiol.* **107**: 1195-1199.

Rabotti G, Zocchi, G. (1994). Plasma membrane bound H⁺- ATPase and reductase activities in Fe-deficient cucumber roots. *Physiol. Plant* **90**: 779-785.

Rae TD, Schmidt PJ, Pufabl RA, Culotta VC, O'Halloran TV (1999). Undetectable intracellular free copper : The requirement of a copper chaperone for superoxide dismutase. *Science* **284**: 805-808.

Rahimi A, Bussler W (1973a). Die Diagnose des Kupfermangels mittels sichtbarer symptome an hoheren Pflanzen. *Landwirtsch. Forsch. Sonderh.* **25**: 42-47.

Rahimi A, Schropp A (1984). Carbonhydraseaktivität und extrahierbares Zink als MaBstab für die Zink-Versorgung von Pflanzen. *Z. Pflanzenernähr. Bodenk.* **147**: 572-583.

Rahimi, A. Bussler W. (1973b). Physiologische Voraussetzungen für die Bildung der Kupfermangelsymptome. *Z. Pflanzenernahr. Bodenk.* **136**: 25-32.

Raineri AM, Cartagna A, Baldan B, Soldatini GF (2001). Iron deficiency affects peroxidase isoforms in sunflower. *J. Exp. Bot.* **52**: 25-35.

Rajaratnam JA, Lowry JB, Avadhani PN, Corley PHV (1971). Boron : Possible role in plant metabolism. *Science* **172**: 1142-1143.

Rajgopalan KV, Johnson JL (1992). The pterin molybdenum cofactor. *J. Biol. Chem.* **267**: 10199-10202.

Rajratanam JA, Lowry JB (1974). The role of boron in the oil palm (*Elaeis guineenis*). *Ann. Bot.* **38**: 193-200.

Ramani S, Kannan S (1985). Studies on Zn uptake and influence of Zn and Fe on chlorophyll development in young maize cultivars. *J. Plant Nutr.* **8**: 1183-1189.

Ramon AM, Carpina-Ruiz RO,Garate A (1989). *In-vitro* stabilization and distribution of nitrate reductase in tomato plants: Incidence of boron deficiency. *J. Plant. Physiol.* **135**: 126-128.

Randall PJ (1969). Changes in nitrate and nitrate reductase levels on restoration of molybdenum to molybdenum deficient plants. *Aust. J. Agric. Res.* **20**: 635-642.

Randall PJ, Bouma D (1973). Zinc deficiency, carbonic anhydrase and photosynthesis in leaves of spinach. *Plant Physiol.* **52**: 229-232.

Rashid A, Bulthio N, Rafiqne E (1994). Diagnosing zinc deficiency in rapeseed and mustard by seed analysis. *Commun. Soil Sci. Plant Anal.* **25**: 3405-3412.

Rashid A, Fox RL (1992). Evaluating internal zinc requirement of grain crops by seed analysis. *Agron J.* **84**: 469-472.

Rathinasabapathy B, Burnet M, Russel BL, Gage DA, Liao P-C, Nye GJ,Scott P, Golbeck JH, Hanson AD (1971). Choline monooxygenase, an unusual iron sulphur enzyme catalyzing the first step of glycine / betaine synthesis in plants. *Proc. Natl. Acad. Sci.* USA **94**: 3454-3458.

Ratnam, K, Shiraishi N, Campbell WH, Hille R (1995). Spectroscopic and kinetic characterization of the recombinant wild type and C 242 S mutant of the cytochrome b reductase fragment of nitrate reductase. J. *Biol. Chem.* **270**: 24067-24072.

Raven JA (1980). Short-and long-distance transport of boric acid in plants. *New Phytol.* **84**: 231-249.

Ravichandran V, Pathmanabhan G (2004). Studies of nodulation and ethanol producing enzymes in cowpea under flood stress regime. *J. Plant Biol.* **33**: 75-80.

Rawson HM (1996a). The developmental stage during which boron limitation causes sterility in wheat genotypes and the recovery of fertility. *Aust. J. Plant Physiol.* **23**: 709-717.

Rawson HM (1996b). Hypothesis for why sterility occurs in wheat in Asia. In *Sterility in Wheat in Sub-tropical Asia: Extent, Causes, and Solutions.* RM Rawson, KD Subedi (eds) Proc. Workshop. September 1995 Lumle Pokhra, Nepal. ACIAR. Canberra, 132-134

Rawson HM, Noppakoonwong RN (1996). Effect of boron limitation in a combination with changes in temperature, light and humidity on floret fertility in wheat. In *Sterility in Wheat with Changes in Subtropical Asia: Extent, Causes and Solutions.* HM Rawson, DK Subedi (eds) Proc. Workshop. September 1995 Lumle Pokhra, Nepal. ACIAR. Canberra, 85-89.

Rea G, Laurenzi M, Tranquilli E, D'Ovidio R, Federico R, Angelini R (1998). Developmentally and wound regulated expression of the gene encoding a cell wall copper amino oxidase in chickpea seedlings. *FEBS Lett.* **437**: 177-182.

Rea G, Metoui O, Infantino A, Federico R, Angelini R (2002). Copper amine oxidase expression in defense responses to wounding and *Ascochyta radiei* invasion. *Plant Physiol.* **128**: 865-875.

Reid RJ, Brooks JD, Tester MA, Smits FA (1996). The mechanism of zinc uptake in plants. Characterization of low-affinity system. *Planta* **198**: 39-45.

Rengel Z (1995a). Carbonic anhydrase activity in leaves of wheat genotypes differing in zinc deficiency. *J. Plant Physiol.* **147**: 251-256.

Rengel Z (1995b). Sulphydryl groups in root-cell plasma membranes of wheat genotypes differing in Zn efficiency. *Physiol. Plant.* **95**: 604-612.

Rengel Z (1997). Root exudation and microflora populations in rhizosphere of crop genotypes differing in tolerance to micronutrient deficiency. *Plant Soil* **196**: 255-260.

Rengel Z (1999b). Physiological mechanisms underlying differential nutrient efficiency of crop genotypes. In *Mineral Nutrition of Crops: Fundamental Mechanisms and Implications.* Z Rengel (ed.) Haworth Press Binghamton, NY. 227-265.

Rengel Z (ed.) (1999a). *Mineral Nutrition of Crops: Fundamental Mechanisms and Implications.* Haworth Press. Binghamton, NY.

Rengel Z, Graham RD (1995a). Importance of seed Zn content for wheat growth on Zn-deficient soil I Vegetative growth. *Plant Soil* **173**: 259-266.

Rengel Z, Graham RD (1995b). Importance of seed Zn content for wheat growth on Zn-deficient soil II Grain yield. *Plant Soil* **173**: 267-274.

Rengel Z, Graham RD (1995c). Wheat genotypes differ in zinc efficiency when grown in chelate buffered nutrient solution I. Growth. *Plant Soil* **176**: 307-316.

Rengel Z, Graham RD (1996). Uptake of zinc from chelate buffered nutrient solution by wheat genotypes differing in Zn efficiency. *J. Exptl. Biol.* **47**: 217-226.

Rengel Z, Graham RD, Pedler JF (1994). Time course of biosynthesis of phenolics and lignin in roots of wheat genotypes differing in manganese efficiency and resistance to take all fungus. *Ann. Bot.* **74**: 471-477.

Rengel Z, Hawkesford MJ (1997). Biosynthesis of a 34 kDa polypeptide in the root-cell plasma membrane of Zn-efficient wheat genotype increases upon Zn starvation. *Austr. J. Plant Physiol.* **24**: 307-315.

Rengel Z, Romheld V, Marschner H (1998). Uptake of zinc and iron by wheat genotypes differing in zinc efficiency. *J. Plant Physiol.* **152**: 433-438.

Rengel Z, Wheal MS (1997). Kinetics of Zn uptake in wheat is affected by herbicide chlorosulphuron. *J. Exp. Bot.* 48: 935-976.

Renger G (1997). Mechanistic and structural aspects of photosynthetic water oxidation. *Physiol. Plant* **100**: 828-841.

Renger G, Wydrzynski T (1991). The role of manganese in photosynthetic water oxidation. *Biol. Metals.* **4**: 73-80.

Rerkasem B (1996). Boron and plant reproductive development. In *Sterility in Wheat in Sub-tropical Asia: Extent, Causes and Solutions.* HM Rowson, KD Subedi (eds) ICIAR, Canberra: 32-35.

Rerkasem B, Jamjod (1997). Genotypic variation in plant response to low boron and implications for plant breeding. *Plant Soil.* **193**: 169-180.

Rerkasem B, Jamjod S (2004). Boron deficiency in wheat: a review. *Field Crops Res.* **89**: 173-186.

Rerkasem B, Netsangtip R, Lordkaew S, Cheng C (1993). Grain set failure in boron-deficient wheat. *Plant Soil* **155/156**: 309-312.

Reuter DJ, Alston AM, Mc Farlanc JD (1988). Occurrence and correction of manganese deficiency in plants. In: *Manganese in Soils and Plants.* RD Graham, RJ Hannan, NC Uren (eds.) Kluwer Academic Publishers Dordrecht.

Reuter DJ, Robinson JB (eds.)(1986). *Plant Analysis: An Interpretation Manual.* Inkata Press Ltd. Melbourne.

Reuter DJ, Robson AD, Loneragen JF, Tranthim-Fryer DJ (1981). Copper nutrition of subterranean clover (*Trifolium subterranium* L. cv. Seaton Park) II. Effects of copper supply on distribution of copper and the diagnosis of copper deficiency by plant analysis. *Aust. J. Agric. Res.* **32**: 267-282.

Rice EL (1984). *Allelopathy.* Academic Press, Orlando FL.

Robbertse PJ, Lock JJ, Stoffbey E, Coetzer LA (1990). Effect of boron on directionality of pollen tube growth in *Petunia* and *Agapanthus. S. Afr. J. Bot.* **56**: 487-492.

Robertson GA, Loughman BC (1974). Response to boron deficiency : A comparison with the responses produced by chemical methods for retarding root elongation. *New Phytol.* **73**: 821-832.

Robinson NJ, Protor CM, Connolly EL, Guerinot ML (1999). A ferric chelate reductase for iron uptake from soils. *Nature* **397**: 694-697.

Robson AD (ed.) (1993). *Zinc in Soils and Plants.* Kluwer Academic Publishers, Dordrecht.

Robson AD, Hartley RD, Jarvis SC (1981). Effect of copper deficiency on phenolic and other constituents of wheat cell walls. *New Phytol.* **89**: 361-373.

Robson RD, Reuter DJ (1981). Diagnosis of copper deficiency. In *Copper in Soils and Plants.* JF Loneragan, AD Robson, RD Graham (eds), Academic Press, London. 287-312.

Rockel P, Strube F, Rockel A, Wildt J, Kaiser WM (2002). Regulation of nitric oxide (NO) production by plant nitrate reductase *in vivo* and *in vitro. J. Exp. Bot.* **53**: 103-110.

Rodriguez IB, Self TR, Peterson GA, Westfall DG (1999). Sodium bicarbonate-DTPA test for the macro and micronutrient elements in soils. *Commun. Soil Sci. Plant Anal.* **30**: 957-970.

Rogers EE, Guerinot ML (2002). FRD3, a member of the multidrug and toxin efflux family, controls iron deficiency responses in *Arabidopsis. Plant Cell* **14**: 1787-1799.

Rognes SE (1980). Anion regulation of lupin asparagine synthetase: Chloride activation of glutamine-utilizing reaction. *Phytochemestry.* **19**: 2287-2293.

Roldin M, Belver A, Rodriguez-Rosales MP, Ferrol N, Donaire JP (1992). *In vivo* and *in vitro* effects of boron on the plasma membrane proton pump of sunflower roots. *Physiol. Plant.* **84**: 49-54.

Romao MJ, Archer M, Maura I, Moura JJG, Le Gall J, Engh R, Schveider M, Hof P, Huber R (1995). Crystal structure of the xanthine oxidase-related aldehyde oxido-reductase from *D. gigas. Science* **270**: 1170-1176.

Romera FJ, Alcantara E, de la Guardia MD (1992). Role of roots and shoots in the regulation of Fe-efficiency responses in sunflower and cucumber. *Physiol. Plant.* **85**: 141-146.

Romera FJ, Alcántara E, De la Guardia MD (1999). Ethylene production by Fe-deficient roots and its involvement in the regulation of Fe-deficiency stress responses by strategy I plants. *Ann. Bot.* **83**: 51-55.

Römheld V (1991). The role of phytosiderophores in acquisition of iron and other micronutrients in graminaceous species. An ecological approach. *Plant Soil* **130**: 127-134.

Römheld V (2000). The chlorosis paradox: Fe inactivation as a secondary event in chlorotic leaves in grapevine. *J. Plant Nutr.* **23**: 1629-1643.

Römheld V, Kramer D (1983). Relationship between proton efflux and rhizodermal transfer cells by iron deficiency. *Z. Pflanzenphysiol.* **113**: 73-83.

Römheld V, Marschner H (1981a). Rhythmic iron stress reactions in sunflower at sub optimal iron supply. *Physiol. Plant.* **53**: 347-353.

Römheld V, Marschner H (1981b). Iron deficiency stress induced morphological and physiological changes in root tips of sunflower. *Physiol. Plant.* **53**: 354-360.

Romheld V, Marschner H (1986). Evidence for a specific uptake system for iron phytosiderophores in roots of grasses *Plant Physiol.* **80**: 175-180.

Römheld V, Marschner H (1990). Genotypic differences among graminaceous species in release of phytosiderophores and uptake of iron phytosiderophores. *Plant Soil* **123**: 147-153.

Roth-Bejerano N, Itai C (1981). Effect of boron on stomatal opening of epidermal strips of *Commelina communis*. *Physiol. Plant* **52**: 302-304.

Rovira AD, Bawden GD, Foster RC (1993). The significance of rhizosphere mycoflora and mycorrhizas in plant nutrition In: *Encyclopedia of Plant Physiology* (NS) 15A-Inorganic Plant Nutrition. A Läuchli, RL Bieleski (eds) Springer Verlag, Berlin. 61-93

Ruiz J M, Baghour M, Bretones G, Belakbir A, Romero L (1998a). Nitrogen metabolism in tobacco plants (*Nicotiana tabacum* L): Role of boron as a possible regulatory factor. *Int. J. Plant. Sci.* **159**: 121-126.

Ruiz J M, Bretones G, Baghour M, Belakbir A, Romero L (1998b). Relationship between boron and phenolic metabolism in tobacco leaves. *Phytochemistry.* **48**: 269-272.

Saeed M, Woodbridge CG (1981). Effect of boron on carbohydrate and amino acids in grape vines. *Pak. J. Agric. Res.* **2**: 108-114.

Sakaguchi T, Nishizawa NK, Nakanishi H, Yoshimura E, Mori S (1999). The role of potassium in the secretion of mugineic acid family phytosiderophores from iron deficient barley roots. *Plant Soil.* **215**: 221-227.

Sakai H, Medrano LJ, Meyerowitz EM (1995). Role of SUPERMAN in maintaining *Arabidopsis* floral whorl boundaries. *Nature.* **378**: 199-203.

Sakal R, Verma MK, Singh AP, Sinha MK (1988). Relative tolerance of some rice varieties to zinc deficiency in calcareous soil. *J. Ind. Soc. Soil Sci.* **36**: 492-495.

Salami AU, Kenefick DG (1970). Stimulation of growth in zinc-deficient corn seedlings by the addition of tryptophan. *Crop Sci.* **10**: 291-294.

Sancenón V, Puigs, Mira H, Thiele DJ, Peòarrubia L (2003). Identification of a copper transporter family in *Arabidopsis thaliana. Plant Mol. Biol.* **51**: 577-587.

Sandmann G, Malkin R (1983). Iron-sulphur centers and activities of photosynthetic electron transport chain in iron deficient cultures of the blue-green algae *Aphanocapsa. Plant Physiol.* **78**: 724-728.

Sandmann G, Reck H, Kessler E, Böger P (1983). Distribution of plastocyanin and soluble plastidic cytochrome c in various classes of algae. *Arch. Microbiol.* **134**: 23-27.

Sasaki H, Hirose T, Watanabe Y, Oshugi R (1998). Carbonic anhydrase activity and CO_2-transfer resistance in zinc-deficient rice leaves. *Plant Physiol.* **118**: 929-934.

Sattelmacher B, Horst WJ, Becker HC (1994) Factors that contribute to genetic variation for nutrient efficiency of crop plants. *Z. Pflanzenernähr. Badenk.* **157**: 215-224.

Sauer P, Frébortová J, Šebela M, Galuszka P, Jacobsen S, Pec P, Frébort I (2002). Xanthine dehydrogenase of pea seedlings: A member of the plant molybdenum oxidoreductase family. *Plant Physiol. Biochem.* **40**: 393-400.

Saxena MC, Chandel AS (1992). Effect of zinc fertilization on different varieties of soybean (*Glycine max*). *Indian J. Agric. Sci.* **62**: 695-697.

Saxena S, Sharma K, Kumar S, Sand NK, Rao PB (2003). Effect of weed extracts on uptake of P and Zn in wheat varieties. *Allelopathy J.* **11**: 201-216.

Scaife A, Turner M (1983). *Diagnosis of Mineral Disorders of Plants.* Vol. 2 *Vegetables*. HMSO, London.

Schikora A, Schmidt W (2001). Iron stress induced changes in root epideRNAl cell fate are regulated independently from physiological responses to low iron availability. *Plant Physiol.* **125**: 1679-1687.

Schmidke I, Stephan UW (1995). Transfer of metal micronutrients in phloem of castor beans (*Ricinus communis*) seedlings. *Physiol. Plant.* **95**: 147-153.

Schmidt W (1999). Mechanisms and regulation of reduction-based iron uptake in plants. *New Phytol.* **141**: 1-26.

Schmidt W, Tillel J, Schikora A (2000). Role of hormone in induction of iron deficiency responses in *Arabidiopsis* roots. *Plant Physiol.* **122**: 1109-1118.

Schnabl H (1980). Anion metabolism as correlated with volume changes in guard cell protoplasts. *Z. Naturforsch.* **35c**: 621-626.

Scholz G, Becker R, Pich A, Stephan UW (1992). Nicotinamine — A common constituent of strategy I and strategy II of Iron acquisition by plants : a review. *J. Plant Nutr.* **15**: 1647-1665.

Schomburg FM, Bizzell CM, Lee DJ, Zeevaart JAD, Amasino RM (2002). Over expression of a novel class of gibberellin 2 – oxidases decreases gibberellin levels and creates dwarf plants. *Plant Cell* **15**: 151-163.

Schon MK, Blevins DG (1990). Foliar boron applications increase the final number of branches and pods on branches of field-grown soybeans. *Plant Physiol.* **92**: 602-607.

Schon MK, Novacky A, Blevins DG (1990). Boron induces hyperpolerization of sunflower root cell memberanes and increases membrane permeability to K+. *Plant Physiol.* **93**: 566-571.

Schubert KR, Boland MJ (1990). The Ureides. In: *The Biochemistry of Plants* BJ Miflin, PJ Lea (eds). Academic Press, San Diego: 197-282.

Schuler MA, Werck-Reichhart D (2003). Functional genomics of P450s. *Annu. Rev. Plant Biol.* **54**: 629-667.

Schürmann P, Jacquot J-P (2000). Plant thrioredoxin systems revisited. *Annu. Rev. Plant Physiol. Plant Mol. Biol.* **51**: 371-400.

Schütte KH (1967). The influence of boron and copper deficiency upon infection by *Erysiphe gramminis*. D.C. the powdery mildew in wheat var. Kenya. *Plant Soil* **27**: 450-452.

Schutte KH, Matthews M (1969). An anatomical study of copper deficient wheat. *Trans. R.Soc. South Africa*. **38**: 183-200.

Schweisguth A, Maier CV, Fox TC, Rumphs ME (1995). Is the anaerobic stress protein enolase a general stress protein. *Plant Physiol.* **106**: S104 (Abstr.).

Scott JM (1989). Seed coatings and treatments and their effects on plant establishment. *Adv. Agron.* **42**: 43-48.

Seckbach J (1982). Ferreting out the secrets of plant ferrtin — A review. *J. Plant Nutr.* **5**: 369-397.

Seethambaram Y, Rao AN, Das VSR (1985). The levels of carbonic anhydrase and of photorespiratory enzymes under zinc deficiency in *Oryza sativa* L. and *Pennisetum americanum* L. Leebe. *Biochem. Physiol. Pflanzen* **180**: 107-113.

Sekimoto H, Seo M, Kawakami N, Komano T, Desloire S, Liotenberg S, Marion-Poll A, Caboche M, Kamiya Y, Koshiba T (1998). Molecular cloning and characterization of aldehyde oxidases in *Arabidopsis thaliana*. *Plant Cell Physiol.* **39**: 433-442.

Seo M, Akaba S, Oritani T, Delarue M, Bellini C, Caboche M, Koshiba T (1998). Higher activity of an aldehyde oxidase in the auxin-overproducing superroot1 mutant of *Arabidopsis thaliana*. *Plant Physiol.* **116**: 687-693.

Seo M, Koshiba T (2002). The complex regulation of ABA biosynthesis in plants. *Trends Plant Sci.* **7**: 41-48.

Shabed A, Bar-Akiva (1967). Nitrate reductase activity as an indication of molybdenum level and requirement of citrus plants. *Phytochemistry* **6**: 347-350.

Shainberg I, Summer ME, Miller WP, Farina MPW, Panan MA, Fey MV (1989). Use of gypsum on soils: A review. *Adv. Soil Sci.* **9**: 1-111.

Sharma CP (ed.) (1996). *Deficiency and Critical Concentrations of Micronutrients in Crop Plants*. Micronutrients Project (ICAR) Bulletin No. 1, Lucknow University Centre, Lucknow.

Sharma CP, Abidi A (1986). Effect of variation in boron supply on growth and metabolism of sunflower. *J. Indian Bot. Soc.* **65**: 471-476.

Sharma CP, Agarwala SC, Chatterjee C, Srivastava SC (1986). Changes in sugarcane metabolism associated with variation in manganese supply. *Sugarcane* **1**: 10-15.

Sharma CP, Agarwala SC, Sharma PN, Ahmed S (1971). Performance of eight high-yielding varieties of wheat in some zinc deficient soils of Uttar Pradesh and their response to zinc amendment in pot culture. *J. Indian Soc. Soil Sci.* **19**: 93-100.

Sharma CP, Chatterjee, C (1997). Molybdenum availability in alkaline soils. In *Molybdenum in Agriculture*. UC Gupta (ed.) Cambridge University Press, New York: 131-149.

Sharma CP, Gupta JP, Agarwala SC (1981). Metabolic changes in *Citrullus* subjected to zinc stress. *J. Plant Nutr.* **3**: 337-344.

Sharma CP, Khurana N, Chatterjee C (1995a). Manganese stress changes physiology and oil content of linseed *Linum usitatissimum* L. *Indian J. Exp. Biol.* **33**: 701-704.

Sharma CP, Mehrotra SC, Sharma PN, Bisht SS (1984a). Water stress induced by zinc deficincy in cabbage. *Curr. Sci.* **53**: 44-45.

Sharma CP, Sharma PN (1987). Mineral nutrient deficiencies affect plant water relations. *J. Plant Nutr.* **10**: 1637-1643.

Sharma CP, Sharma PN, Bisht SS (1984a). Induction of water stress by boron deficiency. *Proc. VIth International Colloquium for the Optimization of Plant Nutrition* Vol. 4. P Martin-Prevel (ed.) IAOPN/GERDT Montpellier: 1303-1307.

Sharma CP, Sharma PN, Bisht SS, Nautiyal BD (1982). Zinc deficiency induced changes in cabbage; In *Proc. 9th Intern. Plant Nutrition Colloquium* Vol. 2. A. Scaife (ed.), Commonw. Agric. Bureaux. Farnham Royal UK. 601-606.

Sharma CP, Sharma PN, Chatterjee C, Agarwala SC (1991). Manganese deficiency in maize affects pollen viability. *Plant Soil* **138**: 139-142

Sharma PN (1992). Pollen fertility in manganese deficient wheat. *Tropical Agric.* **69**: 21-24.

Sharma PN, Mehrotra SC (1998). Diurnal rhythms in leaf Fe in cauliflower grown with deficient and sufficient Fe supply in sand culture. *Ind. J. Exp. Biol.* 36: 924-929.

Sharma PN, Agarwala SC, Sharma CP (1979 a). Boron requirement of hybrid maize var. Ganga 101.In: *Micronutrients in Agriculture*. SC Agarwala, CP Sharma (eds) University of Lucknow, Lucknow. 197-209.

Sharma PN, Chatterjee C, Agarwala SC, Sharma CP (1990). Zinc deficiency and pollen fertility in maize (*Zea mays*). *Plant Soil*. **124**: 221-225.

Sharma PN, Chatterjee C, Sharma CP, Agarwala SC (1987). Zinc deficiency and anther development in maize. *Plant Cell Physiol.* **28**: 11-18.

Sharma PN, Chatterjee C, Sharma CP, Nautiyal N, Agarwala SC (1979b). Effect of zinc deficiency on the development and physiology of wheat pollen. *J. Indian Bot. Soc.* **58**: 330-334.

Sharma PN, Kumar N, Bisht SS (1994). Effect of zinc deficiency on chlorophyll contents, photosynthesis and water relations of cauliflower plants. *Photosynthetica* **30**: 353-359.

Sharma PN, Kumar N, Bisht SS (1996). Guard cell carbonic anhydrase activity and stomatal opening in zinc deficient faba bean, *Vicia faba*. *Ind. J. Exptl. Biol.* **34**: 560-564.

Sharma PN, Ramchandra T (1990). Water relations and photosynthesis in mustard plants subjected to boron deficiency. *Indian J. Plant Physiol.* **33**: 150-154.

Sharma S, Sanwal GG (1992). Effect of Fe deficiency on the photosynthetic system of maize. *J. Plant Physiol.* **140**: 527-530.

Sharma, C.P., Gupta, M. and Pandey, N. (1995b). Zinc deficiency effect on seed coat topography of *Vicia faba* L. cv. Jawahar and VH-130. *J. Indian Bot. Soc.* **74**: 189-192.

Sharma, PN, Tripathi A, Bisht SS (1995c). Zinc requirement for stomatal opening in cauliflower. *Plant Physiol.* **107**: 751-756.

Sharma, UC, Grewal, JS (1990). Potato response to zinc as influenced by genetic variability. *J. Indian Potato Assoc.* **17**: 1-5.

Shelp BJ, Ireland RJ (1985). Ureide metabolism in leaves of nitrogen fixing soybean plants. *Plant Physiol.* **77**: 779-783.

Shelp BJ, Marentes E, Kitheka AM, Vivekanandan P (1995). Boron mobility in plants. *Physiol. Plant.* **94**: 156-161.

Shen Z, Zhang X, Wang Z, Shen K (1994). On the relationship between boron nutrition and development of anther (pollen) in rapeseed plant. *Scientia Agric. Sinica.* **27**: 51-56.

Shen ZG, Liang YC, Shen K (1993). Effect of boron on the nitrate reductase-activity in oilseed rape plants. *J. Plant Nutr.* **16**: 1229-1239.

Sherrell CG (1983). Effect of boron application on seed production of New Zealand herbage legumes. *New Zealand. J. Exp. Agric.* **11**: 113-117.

Shetty AS, Miller GW (1966). Influence of iron chlorosis on pigment and protein metabolism in leaves of *Nicotiana tabacum* L. *Plant Physiol.* **41**: 415-421.

Shetty AS, Miller GW (1966). Influence of iron chlorosis on pigment and protein metabolism in leaves of *Nicotiana tabacum* L. *Plant Physiol.* **41**: 415-421.

Shigaki T, Pittman JK, Hirschi KD (2003). Manganese specificity determinants in the *Arabidopsis* metal/H^+ antiporter CAX2. *J. Biol. Chem.* **278**: 6610-6617.

Shikanai T, Müller-Moulé P, Munekage Y, Niyogi KK, Pilon M (2003). PAA1, P-Type ATPase of *Arabidopsis* functions in copper transport in chloroplasts. *Plant Cell.* **15**: 1333-1346.

Shkolnik MY (1984). *Trace Elements in Plants.* Elsevier Science Publishers, Amsterdam.

Shkolnik MY, Kerpnikova TA, Smirnov YS (1981). Activity of polyphenol oxidase and sensitivity to boron deficiency in monocots and dicots. *Sov. Plant Physiol.* (Engl. Transl.) **28**: 279-283.

Shojima S, Nishizawa N-K, Fushiya S, Nozoe S, Irifune T, Mori S (1990). Biosynthesis of phytosiderophores. *In vitro* biosynthesis of 2-deoxymugineic acid from L-methionine and nicotinamine. *Plant Physiol.* **93**: 1497-1503.

Shorrocks V (1997). The recurrence and correction of boron deficiency. *Plant Soil* **193**: 121-148.

Shrotri CK, Mohanty P, Rathore VS, Tewari MN (1983). Zinc deficiency limits the photosynthetic enzyme activation in *Zea mays* L. *Biochem. Physiol. Pflanzen.* **178**: 213-217.

Shrotri CK, Rathore VS, Mohanty P (1981). Studies on photosynthetic electron transport, photophosphorylation and CO_2 fixation in zinc deficient leaf cells of *Zea mays* L. *J. Plant Nutr.* **3**: 945-954.

Shrotri CK, Tewari MN, Rathore VS (1978). Morphological and ultrastructural abnormalities in zinc deficient maize chloroplasts. *Plant Biochem. J.* **5**: 89-96.

Shrotri CK, Tewari MN, Rathore VS (1980). Effect of zinc nutrition on sucrose biosynthesis in maize. *Phytochemistry.* **10**: 139-140.

Shu Z-H, Wu WY, Oberly GH (1991). Boron uptake by peach leaf slices. *J. Plant Nutr.* **14**: 867-881.

Siedow JN (1991). Plant lipoxygenase — Structure and function. *Annu. Rev. Plant Physiol. Plant Mol. Biol.* **42**: 145-188.

Siedow JN, Umbach AL, Moore AL (1995). The active site of the cyanide resistant oxidase from plant mitochondria contains a binuclear iron center. *FEBS Lett.* **362**: 10-14.

Sigel A, Sigel H (2002). *Molybdenum and Tungsten.* Their roles *in biological processes. Metal Ions in Biological Systems.* Marcel Dekker, New York.

Sijmons PC, Kolattukudy PE, Bienfait HF (1985). Iron deficiency decreases suberization in bean roots through a decrease in suberin-specific peroxidase activity. *Plant Physiol.* **78**: 115-120.

Sillanpaä M (1982). *Micronutrients and the Nutrient Status of Soils.* A global study. Soils Bulletin No. 48. FAO, Rome.

Silverman DN (1991). The catalytic mechanism of carbonic anhydrase. *Can. J. Bot.* **69**: 1070-1078.

Simons K, Ikonen E (1997). Functional rafts in cell membranes. *Nature.* **387**: 569-572.

Sims JL, Eivazi F (1997). Testing for molybdenum availability in soils. In *Molybdenum in Agriculture.* UC Gupta (ed.) Cambridge University Press, New York: 111-130.

Sims JT, Johnson GV (1991). Micronutrient soil tests. In *Micronutrients in Agriculture* 2nd ed. JJ Mortvedt (ed.) Soil Sci. Soc. Am. Book Ser. No. 4. Madison WI. 427-476.

Sinclair TR, Vadez V, Chenu K (2003). Ureide accumulation in response to Mn nutrition by eight soybean genotypes with N_2 fixation tolerance to soil drying. *Crop Sci.* **43**: 592-597.

Singh B, Natesan SKA, Singh BK, Usha K (2005). Improving zinc efficiency of cereals under zinc deficiency. *Curr. Sci.* **88**: 36-53.

Singh K, Chino M, Nishizawa NK, Ohata T, Mori S (1993). Genotypic variation among Indian graminaceous species with respect to phytosiderophore secretion. In *Genetic Aspects of Plant Mineral Nutrition.* PJ Randall, E. Delhaize, RA Richards, R. Munns (eds) Kluwer Academic Publishers, Dordrecht. 335-339.

Singh MV (2001). Evaluation of current micronutrient stocks in different agro-ecological zones of India for sustainable crop production. *Fert. News* **58**: 25-28, 31-38, 41-42.

Singh, M (1981). Effects of zinc, and phosphorus and nitrogen on tryptophan concentration in rice grown on unlimed soils. *Plant Soil.* **62**: 305-308.

Sinha P Sharma CP, Chatterjee C (1999). Seed quality of sesame (*Sesamum indicum*) as influenced by boron. *Indian J. Agric Sci.* **69**: 14-16.

Skerrett M, TyeRNAn SD (1994). A channel that allows inwardly directed fluxes of anions in protoplast derived from wheat roots. *Planta* **192**: 295-305.

Skoog F (1940). Relationship between zinc and auxin in growth of higher plants. *Am. J. Bot.* **27**: 939-950.

Slooten L, Capiau K, Van Camp W, Van Montagu M, Sybesma C, Inze D (1995). Factors affecting the enhancement of oxidative stress tolerance in transgenic tobacco overexpressing manganese superoxide dismutase in the chloroplasts. *Plant Physiol.* **107**: 735-750.

Smith BN (1984). Iron in higher plants : storage and metabolic rate. *J. Plant Nutr.* **7**: 759-766.

Smith GS, Cornforth IS, Henderson HV (1984). Iron requirement of C_3 and C_4 plants. *New Phytol.* **97**: 543-556.

Smith MA, Jonsson L, Stymne S (1992). Evidence for cytochrome b_5 as an electron donor in recinoleic acid biosynthesis in microsomal preperation from developing castor bean (*Ricinus communis* L.) *Biochem J.* **287**: 141-144.

Smith TE, Hetherington SE, Ascher CJ, Stephenson RA (1997). Boron deficiency of avocado 1. Effect on pollen viability and fruit set. In *Proc. Intl. Symp. on Boron in soils and Plants.* RW Bell, B Rerkasem (eds). Kluwer Academic, Dordrecht..

Snir I (1983). Carbonic anhydrase activity as an indicator of Zn deficiency in pecan leaves. *Plant Soil* **74**: 287-289.

Soltanpour PN (1985). Use of ammonium bicarbonate – DTPA soil test elemental availability and toxicity. *Commun. Soil. Sci. Plant Anal.* **16**: 323-388.

Soltanpour PN (1991). Determination of nutrient availability and elemental toxicity by AB-DTPA soil test and ICPS. In *Advances in Soil Science* Vol. **16** BA Stewart (ed.) Springer-Verlag, Berlin: 165-190.

Soltanpour PN, Delgado JA (2002). Profitable and sustainable soil test-based nutrient management. *Commun. Soil Sci. Plant Anal.* **33**: 2557-2583.

Soltanpour PN, Schwab AP (1977). A new soil test for simultaneous extraction of macro-and micro-nutrients in alkaline soils. *Commun. Soil Sci. Plant Anal.* **8**: 195-207.

Somers JJ and Shive JW (1942). The iron-manganese relation in plant metabolism. *Plant Physiol.* **17**: 562-602.

Spencer D (1954). The effect of molybdate on the activity of tomato acid phosphatase. *Aust. J. Biol. Sci.* **7**: 151-160.

Spencer D, Possingham JV (1960). The effect of nutrient deficiencies on the Hill reaction of isolated chloroplasts from tomato. *Aust. J. Biol. Sci.* **13**: 441-445.

Spiller S, Terry N. (1980). Limiting factors in photosynthesis II. Iron stress diminishes photochemical capacity by reducing the number of photosynthetic units. *Plant Physiol.* **65**: 121-125.

Spiller SC, Castelfranco AM, Castelfranco PA (1982). Effects of iron and oxygen on chlorophyll biosynthesis. I. *In vitro* observations on iron and oxygen deficient plants? *Plant Physiol.* **69**: 107-111.

Spiller SC, Kaufman LS, Thompson WF, Briggs WR (1987). Specific mRNA and RNA levels in greening pea leaves during recovery from iron stress. *Plant Physiol.* **84**: 409-414.

Sprague HB (ed.) (1964). *Hunger Signs in Crops—A symposium.* 3rd ed. David Mckay Company, New York.

Srivastava NK, Farooqi AHA, Bansal RP (1985). Response of opium poppy (*Papaver sommniferum*) to varying concentrations of boron in sand culture. *Indian J. Plant Nutr.* **4**: 91-94.

Srivastava NK, Luthra R (1994). Relationship between photosynthetic carbon metabolism and essential oil biogenesis in peppermint under Mn stress. *J. Exp. Bot.* **45**: 1127-1132.

Srivastava NK, Mishra A, Sharma S (1997). Effect of Zn deficiency on net photosynthetic rate, ^{14}C partitioning and oil accumulation in leaves of peppermint. *Photosynthetica* **33**: 71-79.

Steinberg RA, Specht AW, Roller WM (1955). Effect of micronutrient deficiencies on mineral composition, nitrogen fractions, ascorbic acid and the burn of tobacco grown to flowering in water culture. *Plant Physiol.* **30**: 123-129.

Stephan UW, Schmidke I, Stephan U, Scholz G (1996). The nicotianamine molecule is made-to-measure for complexation of metal micronutrients in plants. *Biometals* **9**: 84-90.

Stephan UW, Scholz G (1993). Nicotianmine : mediator of transport of iron and heavy metals in the phloem. *Physiol. Plant.* **88**: 522-529.

Stevens WB, Jolly VD, Hausen NC, Fairbanks DJ (1993). Modified procedures for commercial adaptation of root iron-reducing capacity as a screening technique. *J. Plant Nutr.* **16**: 2507-2517.

Stiefel EI (1996). Molybdenum holsters the bioinorganic brigade. *Science* **273**: 1599-1600.

Storey R (1995). Salt tolerance, ion relations and the effect of root medium on the response of citrus to salinity. *Aust. J. Plant Physiol.* **22**: 101-114.

Strater N, Klabunde T, Tucker P, Witzel H, Krebs B (1995). Crystal structure of a purple acid phosphatase containing a dinuclear Fe (III) - Zn (II) active site. *Science* 268: 1489-1492.

Subedi KD, Gregory PJ, Summerfield RJ, Gooding MJ (1998). Cold temperatures and boron deficiency caused grain set failure in spring wheat *(Triticum aestivum L) Field Crops Res.***57**: 277-288.

Sun T-P, Gubler F (2004). Molecular mechanism of gibberellin signaling in plants. *Annu. Rev. Plant Biol.* **55**: 197-224.

Susin S, Abadia A, Gonzälez-Reyes JA, Lucena JJ, Abadia J (1999). The pH requirement for in vivo activity of the iron-deficiency-induced 'Turbo' ferric chelate reductase. A comparison of the iron deficiency-induced iron reductase activities of intact plants and isolated plasma membrane fractions in sugarbeet. *Plant Physiol.* **110**: 111-123.

Suzuki K, Itai R, Suzuki K, Nakanishi H, Nishizawa N-K, Yoshimura E, Mori S (1998). Formate dehydrogenase, an enzyme of anaerobic metabolism, is induced by iron deficiency in barley roots. *Plant Physiol.* **116**: 725-732.

Syworotkin GS (1958). The boron content of plants with a latex system. Spurenelemente in der Landwirtschaft. Akademie-Verlag, Berlin: 283-288. (in German).

Takagi, S (1976). Naturally occurring iron-chelating compounds in oat and rice root-washings. *Soil Sci. Plant Nutr.* **22**: 423-433.

Takahashi M, Nakanishi H, Kawasaki S, Nishizawa NK, Mori S (2001). Enhanced tolerance of rice to low iron availability in alkaline soils using barley nicotianamine aminotransferase genes. *Nat. Biotechnol.* **19**: 466-469.

Takahashi M, Terada Y, Nakai I, Nakanishi H, Yoshimura E, Mori S (2003). Role of nicotianamine in the intracellular delivery of metals and plant reproductive development. *Plant Cell* **15**: 1263-1280.

Takahashi M, Terada Y, Nakai I, Nakanishi H, Yoshimura E, Mori S, Nishizawa NK (2003). Role of nicotinamine in the intracellular delivery of metals and plant reproductive development. *Plant Cell.* **15**: 1263-1280.

Takahashi M, Yamaguchi H, Nakanishi H, Kanazawa K, Shioin T et al. (1997). Cloning and sequencing of nicotianamine aminotransferase gene (*naat*)–A key enzyme for the synthesis of mugineic acid family phytosiderophores. *9th Int. Symp. Iron Nutr. Interact. Plants.* Stuttgart.

Takahashi M, Yamaguchi H, Nakanishi H, Shioin T, Nishizawa NK, Mori S (1999). Cloning two genes for nicotianamino-transferase, a critical enzyme in iron acquisition (strategy II) in graminaceous plants. *Plant Physiol.* **121**: 947-956.

Takatsuji H (1999). Zinc finger proteins: The classical zinc finger emerges in contemporary plant science. *Plant Mol. Biol.* **39**: 1073-1078.

Takatsuji H, Mori M, Benfey PN, Ren L, Cua NH (1992). Characterization of zinc finger DNA-binding protein expressed specifically in petunia petals and seedlings. *EMBO. J.* **11**: 241-249.

Takkar PN (1993). Requirement and response of crop cultivars to micronutrients in India- a review. In *Genetic Aspects of Plant Mineral Nutrition.* PJ Randall, E Delhaize, RA Richards, R Munns. (eds) Kluwer Academic Publishers, Dordrecht. 341-348.

Takkar PN, Singh SP, Bansal RL, Nayyar VK (1983). Tolerance of barley varieties to Zn deficiency. *Indian J. Agric. Sci.* **53**: 971-972.

Takkar PN, Walker CD (1993). The distribution and correction of zinc deficiency. In *Zinc in Soils and Plants.* AD Robson (ed.) Kluwer Academic Publishers, Dordrecht. 151-165.

Tanada T (1983). Localization of boron in membranes. *J. Plant Nutr.* **6**: 743-749.

Tanaka H (1966). Response of *Lemna pausicostata* to boron as affected by light intensity. *Plant Soil* 25: 425-434.

Tanaka KS, Takio S, Satoh T (1995). Inactivation of Cu/Zn superoxide dismutase induced by copper deficiency in suspension cultured. *Marchamtia paleacea* var. doptera. *J. Plant Physiol.* **146**: 361-365.

Tang C, Robson AD, Dilworth MJ (1990). A split-root experiment shows that iron is required for nodule initiation in *Lupinus angustifolius* L. *New Phytol.* **115**: 61-67.

Tang C, Robson AD, Dilworth MJ (1991). Inadequate iron supply and high bicarbonate impair the symbiosis of peanut (*Arachis hypogaea* L.) with different *Bradyrhizobium* strains. *Plant Soil.* **138**: 159-168.

Tang C, Robson AD, Dilworth MJ, Kuo J (1992). Microscopic evidence on how iron deficiency limits nodule initiation in *Lupinus angustifolius. New Phytol.* **121**: 457-467.

Tang PM, Dela Feunte RK (1986). The transport of indole-3-acetic acid in boron-and calcium-deficient sunflower hypocotyl segments. *Plant Physiol.* **81**: 646-650.

Taniguchi I, Miyahara A, Iwakiri K-I, Hirakawa Y, Hayashi K, Nishiyama K, Akashi T, Hase T (1997). Electrochemical study of biological functions of particular evolutionary conserved aminoacid residues using mutated molecules of maize ferredoxin. *Chem. Lett.* **96**: 929-930.

Tanner PD (1978). The relationship between premature sprouting on the cob and the molybdenum and nitrogen status of maize grain. *Plant Soil* **49**: 427-432.

Taylor IB (1991). Genetics of ABA syntheses In *Abscisic acid, Physiology and Biochemistry.* WJ Davis, HG Jones (eds) Bios Publishers, Oxford 23-37.

Terry N (1977). Photosynthesis, growth and role of chloride. *Plant Physiol.* **60**: 69-75.

Terry N (1980). Limiting factors in photosynthesis I. Use of iron stress to control photochemical capacity in vivo. *Plant Physiol.* **65**: 114-120.

Terry N (1983). Limiting factors in photosynthetic IV. Iron stress mediated changes in light-harvesting and electron transport capacity and its effect on photosynthesis *in vivo*. *Plant Physiol.* **71**: 855-860.

Terry N, Low G (1982). Leaf chlorophyll content and its relation to the intracellular location of iron. *J. Plant Nutr.* **5**: 301-310.

Terry N, Ulrich A (1974). Photosynthetic and respiratory CO_2 exchange of sugarbeet leaves as influenced by manganese deficiency. *Crop Sci.* **14**: 502-504.

Theil EC (1987). Ferritin: structure, gene regulation, and cellular function in animals, plants and microorganisms. *Annu. Rev. Biochem.* 56:289-315.

Thiel H, Finck A (1973). Ermittlung von Grenzwerten optimaler Kupfer Versorgung für Hafer und Sommergertse. *Z. Pflanzenernähr. Bodenk.* **134**: 107-125.

Thomine S, Lelièvre F, Debarbieux E, Schroeder JI, Barbier–Brygoo H (2003). At NRAMP3, a multispecific vacuoler metal transporter involved in plant responses to iron deficiency. *Plant J.* **34**: 685-695.

Thomine S, Wang R, Ward JM, Crawford NM, Schrocder JI (2000). Cadmium and iron transport by members of a plant metal transporter family in *Arabidopsis* with homology to *Nramp* genes. *Proc. Natl. Acad. Sci. USA* **97**: 4991-4996.

Thompson WW, Weier TE (1962). The fine structure of chloroplasts from mineral deficient leaves of *Phaseolus vulgaris. Am. J. Bot.* **49**: 1047-1055.

Thornton RH (1953). *Aspergillus niger* and the estimation of molybdenum availability in soil. *New Zealand Soil News.*3: 42-44.

Tiffin LO (1966). Iron translocation I. Plant culture, exudate sampling, iron citrate analysis. *Plant Physiol.* **45**: 280-283.

Tiffin LO (1972). Translocation of micronutrients in plants. In *Micronutrients in Agriculture.* JJ Mortvedt, PM Giordano, WL Lindsey (eds) Soil. Sci. Soc. Amer. Madison, WI: 199-229.

Timomin ME (1965) Interaction of higher plants and soil microorganisms. In *Microbiology and Soil Fertility.* CM Gilmore, ON Allen, OR Cornalis (eds) Oregon State Univ. Press: 135-138.

Tipton CL, Thowsen J (1985). Fe^{III} reduction in cell walls of soybean roots. *Plant Physiol.* **79**: 432-435.

Toulon V, Sentenac H Thibaud J-B, Davidson J-C, Montineaz, Grigon C (1992). Role of apoplast acidification by the H$^+$ pumps. Effect on the sensitivity to pH and CO$_2$ of iron reduction by roots of *Brassica napus* L. *Planta.* **186**: 212-218.

Treeby M, Marschner H, Romheld V (1989). Mobilization of iron and other micronutrient cations from a calcareous soil by plant-borne, microbial and synthetic metal chelators. *Plant Soil.* **114**: 217-226.

Tsui C (1948). The role of zinc in auxin synthesis in the tomato plant. *Am. J. Bot.* **35**: 172-179.

Tyerman SD (1992). Anion channels in plants. *Annu. Rev. Plant Physiol. Plant Mol. Biol.* **43**: 351-373.

Udvardi MK, Day DA (1997). Metabolic transport across symbiotic membranes of legume nodules. *Annu. Rev. Plant Physiol. Plant Mol. Biol.* **48**: 493-523.

Ulrich A (1952). Physiological basis for assessing the nutritional requirements of plants. *Annu. Rev. Plant Physiol.* **3**: 207-228.

Ulrich A, Hills FJ (1967). Principles and practice of plant analysis. In *Soil Testing and Plant Analysis*, Part II. GW Hardy (ed.) Soil Sci. Soc. Amer., Special publication No. 2, Madison WI. 11-24.

Ulrich A, Ohki K (1956). Chlorine, bromine and sodium as nutrients for sugarbeet plants. *Plant Physiol.* **31**: 171-181.

Utsonomiya E, Muto S (1993). Carbonic anhydrase in plasma membranes from leaves of C$_3$ and C$_4$ plants. *Plant Physiol.* **88**: 413-419.

Vadez V, Sinclair TR (2002) Sensitivity of N$_2$ fixation traits in soybean cultivar Jackson to manganese. *Crop. Sci.* **42**: 791-796.

Vadez V, Sinclair TR, Serraj R (2000). Asparagine and ureide accumulation in nodules and shoots as feedback inhibitors of N$_2$ fixation in soybean. *Physiol. Plant.* **110**: 215-223.

Valee BL, Auld DS (1990). Zinc coordination, function and the structure of zinc enzymes and other proteins. *Biochemistry* **29**: 5647-5659.

Valee BL, Falchuk KH (1993). The biochemical basis of zinc physiology. *Physiol. Rev.* **73**: 79-118.

Van Camp W, Capiau K, Van Montagu M, Inze D, Slooten L (1996). Enhancement of oxidative stress tolerance in transgenic tobacco plants overproducing Fe-superoxide dismutases in chloroplasts. *Plant Physiol.* **112**: 1703-1714.

Van Der Mark F, Bienfait HF, Van Der Ende H (1983). Variable amounts of translatable ferritin mRNA in bean leaves with various iron contents. *Biochem. Biophys. Res. Commun.* **115**: 463-469.

Van der Zaal BJ, Neuteboom LW, Pinas JE, Chardonnens AN, Schat H, Verkleij JAC, Hooykaas PJJ (1999). Over-expression of a novel *Arabidopsis* gene related to putative zinc – transporter genes from animals can lead to enhanced zinc resistance and accumulation. *Plant Physiol.* **119**: 1047-1055.

Van Ginkel G, Sevanian A (1994). Lipid-peroxidation-induced membrane structural alterations. *Methods in Enzymology* **233**: 273-289.

Vaughan AKF (1977). The relation between the concentration of boron in the reproductive and vegetative organs of maize plants and their development. *Rhod. J. Agric. Res.* **15**: 163-170.

Veljovic-Jovanovic S, Oniki T, Takahama U (1998). Detection of monodehydroascorbic acid radical in sulphite-treated leaves and mechanism of its formation. *Plant Cell Physiol.* **39**: 1203-1208.

Veltrup W (1978). Characteristics of zinc uptake by barley roots, *Physiol, Plant.* **42**: 190-194.

Vert G, Briat J-F, Curic C (2001). *Arabidopsis* IRT2 gene encodes a root periphery iron transporter. *Plant J.* **26**: 181-189.

Vert G, Grotz N, Dedaldechamp F, Gaymard F, Guerinot ML, Briat J-F, Curie C. (2002). IRT1, an *Arabidopsis* transporter essential for iron uptake from the soil and for plant growth. *Plant Cell* **14**: 1223-1233.

Viets jr FG, Lindsay WL (1973). Testing soil for zinc, copper, manganese and iron. In *Soil Testing and Plant Analysis* Rev. Ed. LM Walsh, JD Beaton (eds). Soil Sci. Soc. Amer. Madison, WI: 153-172.

Viets jr FG. (1966). Zinc deficiency in the soil-plant system. In: *Zinc Metabolism.* AS Prasad (ed.) Thomas, Springfield. 90-128.

Von Wiren N, Klair S, Bnsal S, Briat JF, Khodr H, Shioiri T, Leigh RA, Hider RC (1999). Nicotianamine chelates both Fe [III] and Fe[II]: Implications for metal transport in plants. *Plant Physiol.* **119**: 1107-1114.

Von Wiren N, Marschner H, Römheld V (1995). Uptake kinetics of iron-phytosiderophores in two maize genotypes differing in iron efficiency. *Physiol. Plant.* **93**: 611-616.

Von Wiren N, Marschner H, Römheld V (1996). Roots of iron-efficient maize also absorb phytosiderophore-chelated zinc. *Plant Physiol.* **111**: 1119-1125.

Von Wiren N, Mori S, Marschner H, Römheld, V (1994). Iron inefficiency in maize mutant *ys1* (*Zea mays* L cv. Yellow-stripe) is caused by a defect in uptake of iron phytosiderophores. *Plant Physiol.* **106**: 71-77.

Von Wirén N, Römheld V, Morel JL, Guckert A, Marschner H (1993). Influence of microorganisms on iron acquisition in maize. *Soil Biol. Biochem* **25**: 371-376.

Walker CD, Loneragan JF (1981). Effect of copper deficiency on copper and nitrogen concentrations and enzyme activities in aerial parts of vegetative subterranean clover plants. *Ann. Bot.* **47**: 65-73.

Walker CD, Webb J (1981). Copper in plants. Form and behaviour. In: *Copper in Soils and Plants.* JF Loneragan, AD Robson, RD Graham (eds) Academic Press, London. 189-212.

Walker-Simmons M, Kudrna DA, Warner RL (1989). Reduced accumulation of ABA during water stress in a molybdenum cofactor mutant of barley. *Plant Physiol.* **90**: 728-733.

Wallace A (1991). Rational approaches to control of iron deficiency other than breeding and choice of resistant cultivars. *Plant Soil* **130**: 281-288.

Wallace A (1992). Some of the problems concerning iron nutrition of plants after four decades of synthetic chelating agents. *J. Plant Nutr.* **15**: 1487-1508.

Wallace A, Lunt OR (1960). Iron chlorosis in horticultural plants. A review. *Proc. Amer. Soc. Hort. Sci.* **75**: 819-841.

Wallace T (1961). *The Diagnosis of Mineral Deficiencies in Plants by Visual Symptoms. A Colour Atlas and Guide* (3rd ed) H.M. Stationery Office, London.

Walsh LM, Beaton JD (eds.) (1972). *Soil Testing and Plant Analysis.* Soil Sci. Soc. Amer. Madison, WI.

Walter A, Pich A, Scholz G, Marschner H, Römheld V (1995). Effects of iron nutritional status and time of day on concentrations of phytosiderophores and nicotianamine in different root and shoot zones of barley. *J. Plant Nutr.* **18**: 1577-1593.

Walter A, Romheld V, Marschner H, Mori S (1994). Is the release of phytosiderophores in zinc-deficient wheat plants a response to impaired iron utilization? *Physiol. Plant.* **92**: 493-500.

Wang L, Reddy KJ, Munn LC (1994). Comparison of ammonium bicarbonate – DTPA, ammonium carbonate and ammonium oxalate to assess the availability of molybdenum in mine spoils and soils. *Commun. Soil Sci. Plant Anal.* **25**: 523-536.

Wang Q, Lu L, Wu X, Li Y, Lin J (2003). Boron influences pollen germination and pollen tube growth in *Picea meyeri. Tree Physiol.* **23**: 345-351.

Watanabe R, Chorney W, Skok J, Wender S (1964). Effect of boron deficiency on polyphenol production in sunflower. *Phytochemistry* **3**: 391-393.

Waters, BM, Blevins DG, Eide DJ (2002). Characterization of FRO 1, a pea ferric chelate reductase involved in root iron acquisition. *Plant Physiol.* **129**: 85-94.

Weiss, MG (1943). Inheritance and physiology of efficiency in iron utilization in soybeans. *Genetics.* **28**: 253-268.

Welch RM (1995). Micronutrient nutrition of plants. *Crit. Rev. Plant Sci.* **14**: 49-82.

Welch RM, Alloway WH, House WA, Kubota J (1991). Geographic distribution of trace element problems. In *Micronutrients in Agriculture* 2ⁿᵈ ed. JJ Mortvedt, FR Cox, LM Schuman, RM Welch (eds) Soil Sci. Soc. Amer. Book Ser. No. 4. Madison WI. 31-57.

Welch RM, Norvell WA (1993). Growth and nutrient uptake by barley *(Hordeum vulgare* L. cv. Herta): Studies using an N- (2-hydroxyethyl) ethylene dinitrilotriacetic acid-buffered nutrient solution technique. I. Role of zinc in the uptake and root leakage of mineral nutrients. *Plant Physiol.* **101**: 627-631.

Welch RM, Norvell WA ,Schaefer SC Schaff JE, Kochian LV (1993).Induction of iron (III) and copper (II) reduction in pea (*Pisum satvum* L.) roots by Fe and Cu status: does the root cell plasmalemma Fe(III) reductase perform a general role in regulating uptake. *Planta.* **190**: 555-561.

Welch RM, Webb MJ, Loneragan JF (1982). Zinc in membrane function and its role in phosphorus toxicity In *Proc. 9ᵗʰ Plant Nutrition Colloquium*, Warwick, England. A. Scaife. (ed.) Commonw. Agric, Bureaux, Farnham Royal, UK. 710-715.

Weleh RM, Norvell WA (1993). Growth and nutrient uptake by barley (*Hordeum inlae* L cv. Herta): Studies using an N̲ (2–hydroxymethyl) ethylene dinitrilotriacetic acid – buffered nutrient solution technique II. Role of zinc in the uptake and root leakage of mineral nutrients. *Plant Physiol.* **101**: 627-631.

Welkie GW (1996). Iron deficiency stress responses of a chlorosis-susceptible and a chlorosis-resistant cultivar of muskmelon as related to root riboflavin excretion. *J. Plant Nutr.* **19**: 1157-1169.

Welkie GW, Miller GW (1993). Plant iron uptake physiology by nonsiderophore systems. In *Iron Chelation in Plants and Soil Microorganisms.* LL Barton, BC Hemming (eds). Academic Press, San Diego, CA: 345-369.

Wenzel AA, Mehlorn H (1995). Zinc deficiency enhances ozone toxicity in bush bean (*Phaseolus vulgaris* L. cv. Saxa). *J. Exp. Bot.* **46**: 867-872.

Westermann RL (ed.) (1990). *Soil Testing and Plant Analysis* 3ʳᵈ ed. Soil Sci. Soc. Amer. Book Series 3. Madison, WI.

Wetzel CLR, Jenson WA (1992). Studies on pollen maturation in cotton: The storage reserve accumulation phase. *Sex. Plant Rep.* **5**: 117-127.

White JG, Zasoski RH (1999). Maping soil micronutrients. *Field Crops Res.* **60**: 11-26.

White MC, Decker AM, Chaney RL (1981). Metal complexation in xylem fluid I. Chemical composition of tomato and soybean root exudate. *Plant Physiol.* **67**: 301-310.

White PJ, Broadley MR (2001). Chloride in soils and its uptake and movement within the plant: A Review. *Ann. Bot.* **88**: 967-988.

Whitehead DC (1985). Chlorine deficiency in red clover grown in solution culture. *J. Plant Nutr.* **8**: 193-198.

Wilkinson RE, Ohki K (1988). Influence of manganese deficiency and toxicity on isoprenoid synthesis. *Plant Physiol.* **87**: 841-846.

Willing RP, Bashe D, Mascarenhas JP (1988). An analysis of the quantity and diversity of messenger RNAs from pollen and shoots of *Zea mays. Theor. Appl, Genet.* **75**: 751-753.

Wilson DO, Boswell FC, Ohki K, Parker MB, Shuman LM, Jellum MD (1982). Changes in soybean seed oil and protein as influenced by manganese nutrition. *Crop Sci.* **22**: 948-952.

Wilson SB, Nicholas DJD (1967). A cobalt requirement for non-nodulating legumes and for wheat. *Phytochemistry* **6**: 1057-1066.

Winkler RG, Palacco JC, Blevims DG, Randall DD (1985). Enzymatic degradation of allantoate in developing soybeans. *Plant Phynol.* **79**: 878-893.

Winkler, RG, Blevins DG, Polacco JC, Randall DD (1987). Ureide catabolism in soybean. II Pathway of catabolism in intact leaf tissue. *Plant Physiol.* **83**: 583-591.

Winsor G, Adams P (1987). *Diagnosis of Mineral Disorders of Plants.* Vol. 3 *Glasshouse Crops.* Chemical Publishing, New York.

Witt HH, Jungk A (1977). Beurteilung der Molybdänversorgung von Pflanzen mit Hilfe der Mo-induzierbaren Nitratreduktase-Activitat. *Z. Pflanzenernähr Bodenk.* **140**: 209-222.

Wittwer SH, Teubner FG (1959). Foliar absorption of mineral nutrients. *Annu. Rev. Plant Physiol.* **10**: 13-32.

Wood JG, Sibly PM (1952). Carbonic anlydrase activity in plants in relation to zinc content. *Aust. J. Sci. Res.* **(B)5**: 244-255.

Woods WG (1996). Review of possible boron speciation relating to essentiality. *J. Trace Elem. Exp. Med.* **9**: 153-163.

Wu G, Wilen RW, Robertson AJ, Gusta LV (1999). Isolation, chromosomal localization, and differential expression of mitochondrial manganese superoxide dismutase and chloroplastic copper/zinc superoxide dismutase genes in wheat. *Plant Physiol.* **120**: 513-520.

Wu Z, Liang F, Hong B, Yaung JC, Sussman MR, Harper JF, Sze H (2002). An endoplasmic reticulum – bound Ca^{2+}/Mn^{2+} pump ECA 1, supports plant growth and confers tolerance to Mn^{2+} stress. *Plant Physiol,* **130**: 128-137.

Xu G, Magen H, Tarchitzky J, Kafkafi U (2000). Advances in chloride nutrition. *Adv. Agron.* **66**: 96-150.

Xu H, Huang Q, Shen K, Sten Z (1993). Anatomical studies on the effects of boron on the development of stamen and pistil of rape (*Brassica napus* L.). *Zhiwa Xuebao* **35**: 453-457.

Xu H. (1993). The effects of boron deficiency on the floral anatomy of oil seed rape (*Brassica napus*). *Acta Botanica Sinica* **35**: 453-457.

Yachandra VK, De Rose VJ, Latimer MJ, Mukerji I, Sauer K, Klein MP (1993). Where plants make oxygen: A structural model for the photosynthetic oxygen-evolving manganese complex. *Science.* **260**: 675-679.

Yamagishi M, Yamamoto Y (1994). Effects of boron on nodule development and symbiotic nitrogen fixation in soybean plants. *Soil. Sci. Plant Nutr.* **40**: 265-274.

Yamaguchi H, Nishizawa N-K, Nakanishi H, Mori S (2002). ID17, a new iron-regulated ABC transporter from barley roots, localizes to the tonoplast. *J. Exp. Bot.* **53**: 727-735.

Yamasaki H, Sakihama Y (2000). Simultaneous production of nitric oxide, peroxinitrite by plant nitrite reductase: *In vitro* evidence for the NR-dependent formation of reactive nitrogen species. *FEBS Letters* **468**: 89-92.

Yamasaki H, Sakihama Y, Takahashi S (1999). Alternative pathways for nitric oxide production in plants: new features for an old enzyme. *Trends Plant Sci.* **4**: 128-129.

Yan F, Schubert S, Mengel, K (1992). Effect of low root medium pH on the net proton release, root respiration and root growth of corn (*Zea mays* L.) and broad beans. *Plant Physiol.* **99**: 415-421.

Yanagisawa S (1995). A novel DNA binding domain that may form a single zinc finger motif. *Nucleic Acids Res.* **23**: 3403-3410.

Yanagisawa S, Sheen J (1998). Involvement of maize Dof zinc–finger proteins in tissue specific and light –regulated gene expression. *Plant cell.* **40**: 75-89.

Yang X, Römheld V, Marschner H, Chaney RL (1994). Application of chelated buffered nutrient solution technique in studies on zinc nutrition in rice plant (*Oryza sativa* L.). *Plant Soil.* **163**: 85-94.

Yi Y, Guerinot ML (1996). Genetic evidence that induction of root Fe(III) chelate reductase activity is necessary for iron uptake under iron deficiency. *Plant J.* **10**: 835-844.

Yip JY, Vanlerberghe GC (2001). Mitochondrial alternative oxidase acts to dampen the generation of active oxygen species during a period of rapid respiration induced to support light rate nutrient of uptake. *Physiol. Plant.* **112**: 327-333.

Youngdahl LJ, Svel LV, Liebhard WC, Tell MR (1977). Changes in ^{65}Zn distribution in corn roots tissue with a phosphorus variable. *Crop Sci.* **17**: 66-69.

Younge OR, Takahashi M (1953). Response of alfalfa to molybdenum in Hawaii. *Agronomy J.* **45**: 420-428.

Yu G, Osborne LD, Rengel Z (1998). Micronutrient deficiency changes activities of superoxide dismutase and ascorbate peroxidase in tobacco plants. *J. Plant Nutr.* **21**: 1427-1437.

Yu Q, Baluska F, Jasper F, Menzel D, Goldbach E (2002). Short term boron deprivation enhances levels of cytoskeletal proteins in maize, but not in zucchini, root apices. *Physiol. Plant.* **117**: 270-278.

Yu Q, Hlavacka A, Matoh T, Volkmann D, Menzel D, Goldbach HE, Baluska F (2003). Short-term boron deprivation inhibits endocytosis of wall pectins in meristematic cells of maize and wheat root apices. *Plant Physiol.* **130**: 415-421.

Yu Q, Rengel Z (1999a). Micronutrient deficiency changes activities of superoxide dismutase and ascorbate peroxidase in tabacco plants. *J. Plant Nutr.* **21**: 1427-1437.

Yu Q, Rengel Z (1999b). Micronutrient deficiency influences plant growth and activities of superoxide dismutases in narrow-leafed lupins. *Ann. Bot.* **83**: 175-182.

Yu Q, Worth C, Rengel Z (1999). Using capillary electrophoresis to measure Cu/Zn-SOD concentration in leaves of wheat genotypes differing in tolerance to Zn deficiency. *Plant Sci.* **143**: 231-239.

Zaharieva T, Romheld V (2001). Specific Fe^{3+} uptake system in strategy I plants inducible under iron deficiency. *J. Plant Nutr.* **23**: 1733-1744.

Zaid H, Morabet R El, Diem HG, Arahou M (2003). Does ethylene mediate cluster root formation under iron deficiency. *Ann.Bot.* **92**: 673-677.

Zhang F, Römheld V and Marschner H (1991a). Release of zinc mobilizing root exudates in different plant species as affected by zinc nutritional status. *J. Plant Nutr.* **14**: 675-686.

Zhang F, Römheld V, Marschner H (1991b). Diurnal rhythm of release of phytosiderophores and uptake rates of zinc in iron-deficient wheat. *Soil Sci. Plant Nutr.* **37**: 671-678.

Zhang F, Römheld V, Marschner H (1991c). Role of the root apoplasm for iron acquisition by wheat plants. *Plant Physiol.* **97**: 1302-1305.

Zhang N, Portis AR (1999) Mechanism of high regulation of Rubisco: A specific role for the larger Rubisco activase isoforms involving reductive activation of thioredoxin-f. *Proc. Natl. Acad. Sci. USA.* **96**: 9438-9443.

Zhang X, Shen Z, Shen K (1994). Effect of boron on floral organ development and seed setting of rape seed (*Brassica napus* L.). *Turang Xuebas* **31**: 146-152.

Zhang, F, Römheld, V Marschner, H (1989). Effect of zinc deficiency in wheat on the release of Zn and Fe mobilizing root exudates. *Z. Pflanzenernacher. Bodenk.* **152**: 205-210

Zheng SJ, Tang C, Arakawa Y, Masaoka Y (2003). The responses of red clover (*Trofolium pratense* L.) to iron deficiency: a root Fe(III) chelate reductase. *Plant Science* **164**: 679-687.

INDEX